The Illustrated Guide to
CACTI

The Illustrated Guide to
CACTI

RUDOLF SLABA

Illustrations by
PETR LISKA

CHANCELLOR
PRESS

Translated by Olga Kuthanová
Graphic design by Eva Adamcová
First published in Great Britain in 1992 by
Chancellor Press,
Michelin House, 81 Fulham Road
London SW3 6RB
ISBN 1 85152 118 6
Printed in Czechoslovakia
3/15/29/51-01

CONTENTS

The Origins of Cacti 6
Land of Origin and Distribution of Cacti 7
Main Bodily Organs 8
Systematic Classification and Nomenclature 12
Main Tasks of the Cactus Grower 13
  Planting and repotting 13
  Watering 15
    Spraying 17
    Watering from above 17
    Watering from below 17
  Propagation 18
    Propagation from cuttings 18
    Propagation from seed 18
    Grafting 20
      Rootstock 21
      Methods of grafting 21
Diseases and their prevention 23
Pests 25
Colour Plates 27
  Cacti of North America 28
  Cacti of South America 110
Index 221

THE ORIGINS OF CACTI

Cacti are without doubt one of the most typical groups of succulents. Their present appearance is the result of a lengthy process of adaptation throughout a long and difficult period of evolution.

It is presumed that the ancestors of the cactus family first appeared on the scene at the end of the Mesozoic and beginning of the Tertiary periods, a time marked by the great development of angiosperms. In response to continually deteriorating climatic conditions, and especially a periodically repeated general decrease in the amount of rainfall, plants throughout tropical South America, growing in habitats that were greatly affected by exposure to the sun, began to develop adaptations that enabled them to withstand temporary periods of drought by storing a supply of water in their stems, which developed to become more fleshy. These predecessors of the cacti did not differ much from other leafy plants at first and only gradually perfected the changes in their body organs that were necessary for succulence. Existing representatives of this family provide us with some idea of its evolution. It is a relatively young, non-stabilized branch from which all the evolutionary stages have not as yet been eliminated and hence its numbers currently include a great many direct or only partially improved descendants of the original, ancestral cacti. Those plants most like the ancestral cacti are the Pereskias (members of the subfamily Pereskioideae), which may be considered the predecessors of the succulent cacti.

The evolution of cacti was a complex process of adaptation, of improved changes in numerous morphological characteristics and an increase in the degree of succulence. The development of areoles is one example. These are modified branches, and the spines they bear are modified leaves. The stem, or body, likewise underwent marked changes, gradually becoming shorter and more globose. Following the uplift of the Andes, however, many cacti found themselves in a more humid environment once again and it is presumed that a great number did not survive these changes and became extinct. Some, however, were able to adapt to the new conditions and developed flatter, broader stems resembling leaves, but not true leaves. Typical representatives of these cacti in the more humid habitats of South America are members of the genus *Schlumbergera* (Christmas Cactus), *Epiphyllum, Rhipsalis* and many others.

Although the American continent did not have its present form at the beginning of the Tertiary period, it was still isolated from the other land masses and the first cacti did not penetrate to other continents, just as many African succulents did not make their way to the Americas. The only possible route the original cacti had for expansion was apparently via the geologically extremely unstable territory of present-day central America and the Antilles, a land passage that occasionally emerged from the sea to join North and South America. With the passage of time, cacti in far distant parts of North and South America evolved along different lines, giving rise to the northern and southern branches of the cactus family (Cactaceae). It is believed that all known species of cacti already existed as far back as the early Quaternary period.

LAND OF ORIGIN AND DISTRIBUTION OF CACTI

Cacti are native to the American continent whence members of certain genera have spread to other parts of the world. Cacti of the genus *Rhipsalis*, for example, were most probably introduced into Africa and India by birds, while some species of *Opuntia* were brought to Europe by people, following the discovery of America, where they became established and partly naturalized as escapees, chiefly in the region of the Mediterranean Sea. Opuntias likewise took possession of, and established themselves as weeds in, vast areas along the coast of Africa, and in Australia to where they were brought in 1825 for use as hedging plants.

In North and South America cacti are found chiefly in warm, dry regions, with practically no limit to their altitudinal distribution. They may be found at sea level as well as high up in the mountains, at elevations of over 4,000 m (13,000 ft). Their greatest concentration, as well as the greatest diversity of species, is in the region along the Tropic of Cancer in North America (Mexico and the southern United States) and in South America along the Tropic of Capricorn (northern Argentina, Paraguay, southern Brazil). Some species occur sporadically as far north as 53° in Canada, and as far south as 50° in the inhospitable land of Patagonia.

MAIN BODILY ORGANS

Cacti belong to the group of dicotyledonous angiospermous plants, although, at first glance, this is evident only in the case of the most primitive cacti. Two seed leaves (cotyledons) are evident primarily in all germinating plants of the subfamilies Pereskioideae and Opuntioideae. In the more advanced genera the seed leaves are already so thickened and fleshy that they are practically unrecognizable as such. They occur as a sort of globose structure divided by a shallow groove into two halves — the body of the cactus to be.

The body of the cactus, referred to as corpus in all languages by cactus growers the world over, is a succulent stem through the centre (axis) of which passes the main vascular bundle, namely from the roots to the crown of the plant, the main vegetative point. Round the vascular bundle there is a layer of parenchymatous tissue of varying thickness (depending on the degree of succulence), which is able to store water. The more advanced cacti are on the evolutionary scale, the thicker are their stems. Because cacti do not possess leaves (with the exception of the subfamily Pereskioideae), their function, i. e. assimilation and transpiration, has had to be taken over by the

Cactus body — longitudinal section:
A plant with continuous, unidivided ribs; B plant with ribs divided into tubercles; 1 crown (chief vegetative centre); 2 main vascular bundle; 3 auxiliary vascular bundles; 4 areole with spines; 5 axil; 6 root neck; 7 root

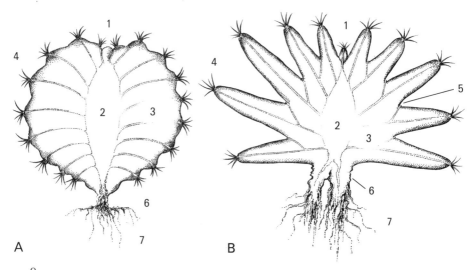

body. To prevent excessive loss of water during dry periods, the outer skin of the stem (epidermis) has only very few pores and is greatly thickened or has a waxy coating. The surface of the stem is sometimes almost smooth (for instance in the genus *Opuntia*) but more generally it is ribbed (the ribs varying in height, width, and roundness) or tuberculate, which permits the plant to shrink in dry periods, thereby enabling it to withstand water loss.

Besides the most common forms — columnar, cylindrical, globose, disc-like, arborescent, candelabrum-like, tufted, pendent and creeping, there are also other rarer, unusual forms, namely cristate and monstrous forms. Whereas in the monstrous forms the stem is greatly increased in bulk, excessively enlarged at the sides, in cristate forms the crown takes the shape of a raised ridge that is fan-shaped at first, and later greatly bent in several places. What caused these growth abnormalities and what purpose they serve has never been satisfactorily explained.

A striking characteristic of cacti is their generative points, called areoles — small pincushion-like structures that bear spines and generally also flowers and shoots. All areoles are linked by auxiliary vascular bundles to the main vascular bundle. If, for some reason, the vegetative point at the crown dies, then a new shoot will grow from one of the areoles. In some cacti (e. g. *Mammillaria*), however, the vegetative points are in the axils of the mamillae or in the grooves joining the areoles and axils (e. g. in *Coryphantha*). The areoles and axils are usually covered with wool or felt, which in some species remains on the plant a long time, whereas in others it soon disappears.

Cacti's greatest attraction, apart from their flowers of course, is their spines. Spines are modified leaves and, according to their arrangement in the areole, they are either central or radial. Sometimes, however, the central spines may be absent or are formed at a later age. A special type of spine, peculiar to the

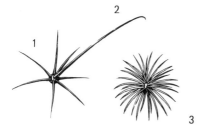

Arrangement of spines in the areole:
1 radial spines; 2 central spines;
3 radiate spines;
4 pectinate spines (arranged like the teeth of a comb)

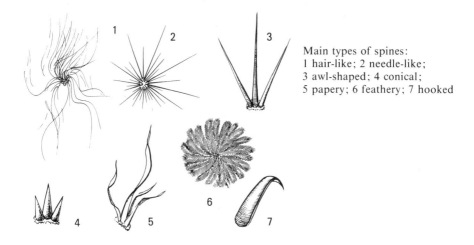

Main types of spines:
1 hair-like; 2 needle-like;
3 awl-shaped; 4 conical;
5 papery; 6 feathery; 7 hooked

Opuntias, for example, are the short, weak spines called glochids. Spines vary widely in shape. A spine consists of a base, middle section and tip. As a rule, each section is a different colour, especially during the period of early growth when the spines are at their most colourful, but also at their softest, so that they can easily break away from the areole. Only later do they gradually dry out and become hard and firm. Spines protect the cacti from animal predators and, in many species, they also shade the skin and thus provide protection against water loss during the resting period.

The point where the body of a cactus joins its root is called the root neck. The root neck is relatively prone to fungus rot and for that reason should be left partly exposed, especially in the case of more delicate species. Roots may be turnip-like tap roots or greatly branched. Cacti with turnip-like tap roots require a heavier, more compact loam, whereas cacti with a much-branched, spreading root system are better off in a light, well-aerated soil.

Parts of a flower:
1 inner perianth segments;
2 outer perianth segments;
3 stigma;
4 style;
5 nectar-secreting chamber (nectary);
6 ovary;
7 pericarp;
8 anthers;
9 filaments (of stamens)

10

Main types of flowers:
1 actinomorphic
(radially symmetrical, regular);
2 zygomorphic
(bilaterally symmetrical)

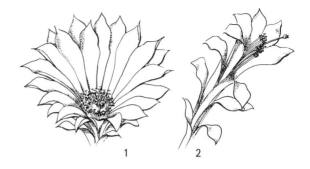

1   2

Cacti's magnificent, albeit transient, ornament is their flowers. These grow from areoles, axils or grooves and are located either on the crown or on the sides of the stem. In some cacti the flowers grow from areoles clustered together in a so-called cephalium. If the cephalium forms at the top of the stem (e. g. in *Melocactus, Discocactus*), it is called a terminal cephalium; if it is on the side (e. g. in *Cephalocereus*), it is called a lateral cephalium.

The flowers of cacti are distinguished by many different aspects. As regards shape they may be dish-like, funnel-shaped, bell-shaped, tubular, etc. In terms of the arrangement of the individual parts, they are either actinomorphic (radially symmetrical) or zygomorphic (bilaterally symmetrical). Flowers of the former type are more common in cacti; only a few genera (e. g. *Cochemiea, Bolivicereus, Matucana*, etc.) have zygomorphic flowers.

The flowers are further distinguished according to their manner of pollination and fertilization and may be either autogamous (pollinated by their own pollen, in other words self-pollinated) or allogamous (reproducing by cross-pollination). Cross-pollination is the prevalent method among the whole cactus family, and thus in practically all instances at least two genetically different individuals are necessary for fertilization. In the wild, diurnal flowers are pollinated mostly by insects. For example, the narrowly long-tubed zygomorphic flowers of *Borzicactus, Oreocereus, Cleistocactus* etc. are pollinated by hummingbirds. Nocturnal white, pleasantly scented flowers, such as those of *Discocactus* and *Cereus*, are pollinated by hawk moths. On the other hand, the flat, dish-like, foetid flowers of *Neobuxbaumia, Carnegiea, Cephalocereus*, etc., are pollinated by bats. Self-pollinated species, in which all that is needed is for the pollen to be transferred from the anthers to the stigma of the same flower, are few in number. Self-polli-

11

nation is common in *Melocactus* and *Frailea,* and occasional in *Notocactus* and *Rebutia.*

Following fertilization, the ovary ripens into a fruit — a berry that contains up to several thousands of seeds. The seeds are fully ripe only when the fruit bursts or dries up, or else falls spontaneously from the areole, axil or cephalium. The fruits of some species, for example *Mammillaria* and *Coryphantha,* remain concealed inside the body of the cactus and only after a lengthy period, sometimes not until the following year, do they appear on the surface.

SYSTEMATIC CLASSIFICATION AND NOMENCLATURE

All cacti belong to the family Cactaceae. The determining character of identification for the whole family is the presence of areoles — a feature possessed only by cacti. Today the family Cactaceae comprises the following three subfamilies: Pereskioideae, Opuntioideae, and Ccrcoideae. The subfamily Pereskioideae comprises the most primitive cati, with normal leaves and flowers arranged in racemes. The leaves of cacti of the subfamily Opuntioideae are practically rudimentary and the areoles are furnished with glochids. The subfamily Cereoideae includes the most advanced forms of cacti, in which the leaves are missing and the body is completely succulent.

We shall bypass the further division of cacti into families, which is not of particular importance in general practice, and turn our attention to the most commonly used taxonomic units, namely genus, species, variety and form. Whereas Linnaeus classed the 21 species known to him in his day in the single genus *Cactus,* Salmus Dyck's, and later Schumann's, system from the late nineteenth century already comprised 21 genera. The American botanists L. N. Britton and J. N. Rose, in a monograph dating from 1919—23, classed cacti in 124 genera, and Backeberg's system from the beginning of the second half of the twentieth century contained 220 genera. Since the time of Linnaeus botanists have tried to come up with a perfect system, but so far without success. According to recent knowledge, it is necessary to combine certain genera for whose existence as a separate genus there is no substantiation from the botanical viewpoint.

We may consider that the family Cactaceae comprises approximately 150 genera with about 2,500 species. It is not possible to give a more exact number because there are differences of opinion as to the taxonomic value of many genera as well as species. At the same time, some species, previously classed as separate, independent species, must now be seen from the botanical viewpoint merely as varieties, forms or even invalid names (synonyms) of previously described plants.

Besides the great number of separate species, individual species occur in numerous varieties and forms. Varieties exhibit marked, permanently hereditary, deviations from the type species; forms exhibit minor deviations that need not be hereditary.

Besides the Latin name of the genus and species, and possibly the variety or form, the botanical name of a cactus also includes the initials or full name of the person who first described the given plant. If, for some reason, the species is later re-classed in a different genus, or if its taxonomic value changes (e. g. it is proved that the described cactus is merely a variety of a previously described species), the name of the person who gave the plant its first name appears in parentheses followed by the name of the person who made the change.

The scope of this book allows for the inclusion of only a small fraction of the vast number of cacti discovered to date. It therefore presents those species that are most widely grown, or that possess some unusual characteristic, and purposely leaves out those species, and even whole genera, that are difficult to grow in conditions available to the amateur. This, of course, means that the cacti in the ensuing pictorial section could not be arranged according to one or another of the most commonly used systems at least partly elucidating the relationships within, and indicating the evolution of, the whole family. The individual genera are arranged in only two groups, according to the area of their distribution. The first includes cacti of North America; the second cacti of South America. In each the genera and species are arranged in alphabetical order.

MAIN TASKS OF THE CACTUS GROWER

Planting and Repotting

Repotting cacti is without doubt one of the most important and time-consuming of the cactus grower's tasks. Plants that are transplanted regularly thrive and flower profusely. Whereas the smallest seedlings should be lifted and reset in fresh soil annually, the interval for older specimens may be three to five years or even longer if they are given applications of liquid feed.

The best containers for growing cacti are plastic pots. These are not only lightweight but, being non-porous, prevent the loss of moisture from the soil, which is instrumental in maintaining the good condition of the root system. Square pots are recommended because they hold more soil and allow for more profuse branching of the roots. However, this does not mean that all cacti will grow better the more soil they have. This applies only to certain vigorously growing species with a large root system, such as *Cereus, Opuntia, Echinopsis,* and the like. For the majority of globose cacti, the width of the container should not exceed the width of the stem by more than one quarter of its diameter.

Most cacti will be satisfied with a container that is as deep as it is wide. Cacti with a prominent turnip-like tap root require a deeper pot so that the main root does not curve upwards when it reaches the bottom of the container.

Cacti are generally repotted in spring or during the resting period in summer, i. e. end of July and beginning of August in western Europe. May and June are the least suitable times, for this is the period of most vigorous growth. If you plan to repot your cacti in winter or early spring, the soil should be prepared in advance in the autumn.

It is most important that the soil mixture is free-draining, loose, friable when squeezed in the hand even when moist, and nutrient-rich, but at the same time without containing too many undecomposed organic particles. A substrate that is fully satisfactory and suitable for practically all species of cacti is a mixture of one part rather light, rich soil and one part coarse river sand. For more demanding species a mixture of equal parts of soil, sand and peat, or soil, sand, peat and crushed brick can be recommended. The more susceptible a cactus is to root rot, for instance many Mexican rarities, the greater should be the proportion of sand and coarse, crushed brick in the substrate. The decrease in the substrate's nutrient value as a result of the greater proportion of inorganic components may be offset by the occasional application of liquid feed. Cacti may likewise be grown successfully in practically sterile substrates (e. g. crushed brick, granite rubble, peat, volcanic rubble, polyurethane mixtures, etc.), food being supplied by watering with nutrient solutions.

As to the actual method of repotting, the grower would be well advised to heed the following instructions. The cactus should not be removed from the pot by force, i. e. pulling or twisting it out. The entire contents of the pot should be loosened either by turning the pot upside down and tapping it against the edge of a table while holding the plant with the other hand, or else by inserting a peg into the drainage hole and pressing the root ball out. Remove all soil particles clinging to the roots, as well as dead rootlets, and, with a sharp knife, remove all roots that are partly decayed, scabby, felted, or infested with pests. Such radical removal of all unhealthy parts is a must, even at the cost of cutting all the roots. The remaining healthy roots should then be shortened to promote branching. Roots that have been damaged during handling should be allowed to dry for several days. The larger the plant, the longer it should be left out in the open. It is recommended that one-year-old seedlings be put in pots after two or three days, while adult specimens should be allowed to dry for at least a week.

A very good method of rooting cacti is by dry rooting. Place the cactus with its shortened roots on top of an empty pot and put it in a warm, shaded place. When the gleaming tips of new rootlets appear on the main roots (usually after about three or four weeks), pot up the plant in soil.

Before putting the plant in a new pot, cover the drainage holes with crocks and top them with a drainage layer of coarse sand. After this, fill the pot to about one third with a rather dry soil mixture. Then put the cactus roots inside the pot and gradually add soil on all sides with a spoon or trowel or simply with your hand until the pot is filled nearly to the rim. See to it that the roots extend downwards, that they are spread out evenly, and that their tips do not curve upwards. Tap the pot lightly in order to settle the soil mixture. If part of the stem is embedded in the soil, carefully pull the cactus up a bit. Last of all, add a thin layer of coarse sand or fine gravel. This serves to prevent rotting of the stem or it becoming infected through contact with the soil and also helps to anchor the cactus in the pot. It likewise limits loss of soil moisture and facilitates the absorption of water when the plant is watered from above.

Watering

Few houseplants can do without water as long as cacti. Even

during the growing season cacti can be left untended for a whole month, for instance when you go on vacation, without any need to worry about their fate. Experiments have shown that cacti can survive for several years without water. However, our aim is not mere survival but the growth of cacti to their desired size, and in this case more frequent watering is naturally a must.

How often cacti should be watered depends on many circumstances. Cacti placed by a window are watered differently from species grown in a rooftop glasshouse, and quite differently again from plants grown in a frame that is partly shaded by trees. Correct application of water also depends on the plant's size. In the case of small seedlings, the substrate should not be allowed to dry out. In the case of adult collection specimens, water should occasionally be withheld even during the growing period, at which time the plants produce beautiful spines and increase in size only slightly, making use of the supply of water stored in their succulent stems. Watering also depends on the size of the container and the composition of the substrate. Plastic boxes dry out much slower than clay pots and a loose, free-draining substrate, with a greater proportion of sand and crushed brick, dries out more quickly. Last but not least, water should be applied in accordance with the types of specimens in the collection. Whereas cacti from habitats with relatively moist conditions (e. g. *Notocactus, Frailea, Melocactus*) should be watered as soon as the soil dries, cacti from dry regions, with a delicate root system (e. g. many Mexican and Chilean cacti), should be watered only after a lengthier period, even as much as several weeks, when the soil is completely dry.

The last, and perhaps most important, factor is the season of the year and the immediate climatic conditions. In winter, water should be withheld altogether except for occasional and careful watering of the smallest seedlings. In spring, which is the period of most active growth, water should be supplied liberally. In hot summer weather, many cacti suspend their growth and hence watering should be limited. In the autumn, cacti practically cease growing but produce the longest and loveliest spines. At this time they should not be watered at all because the soil will remain moist for a long time. If there is a pronounced drop in temperature for a long period during the growing season, water should also be withheld at this time.

Based on the foregoing it is possible to provide growers with the following general rule of thumb. Intervals for watering range from one week (e. g. in the case of small seedlings) to four or more weeks. Water should be supplied liberally so that it penetrates throughout all layers of the substrate. Do not water when the soil is moist, otherwise the roots, and possibly even the whole plant, might die. It is better to supply water liberally and less frequently, then lightly and often. In warm and dry weather, water should be applied more frequently; in cold, rainy weather, water should be withheld altogether.

There are three main methods of watering cacti:

Spraying
Spraying or syringing is the method used mainly with small seedlings, to prevent them from being washed out of the substrate. Cacti are sprayed during the period when they are to begin growth (early spring) and also when they do not absorb water willingly through the roots, which is usually during the resting period in summer and immediately after transplanting. Spraying should also be the method used whenever applying solutions to protect the plants against pests and diseases.

Watering from above
Cacti should be watered from above at the beginning of the growing season (in early spring), during the period of slower growth in summer, and during rooting, in other words when the substrate must not be too wet for too long. In smaller collections, each plant should be watered separately; in larger collections cacti are watered with a watering can or with a spray attachment. The advantage of this method is that besides being quick and assuring optimum moistening of the substrate, it also increases the overall atmospheric moisture. A drawback is that it washes the wool off the areoles and causes delicate, hair-like spines to droop, which, in some species (e. g. *Espostoa*), detracts from their appearance. It should be added that the wool on the areoles soon grows in again, sometimes even to the benefit of the plant's appearance if the water has removed old, dingy wool and dust. Even in their native habitats, cacti are regularly divested by rain of their fine wool, which grows in anew at the end of the rainy season.

Watering from below
In this case, the pots are placed in shallow bowls or basins filled with water that penetrates all layers of the substrate by osmosis. Under optimum growth conditions, particularly at

higher temperatures, it is not necessary to remove excess water from the basins. This method of watering is suitable in spring and late summer, and also when applying fertilizer mixed with water or solutions used to protect the roots from pests. It is unsuitable during the period of root formation in early spring, following transplanting and rooting, or in the autumn when the substrate dries poorly.

Cacti should not be watered with hard water or water that is too cold. Watering from above or syringing should be done in the evening, or else in the morning when the epidermis is not as warm as at other times of the day, and when the flowers are closed. Watering from below should be done during the day.

Propagation

Propagation from cuttings

Cacti are propagated from seed or by vegetative means, i. e. by shoots and cuttings or by grafting. Vegetative propagation is the simplest method and therefore this will be discussed first.

A good time for removing offshoots or taking cuttings is in spring and early summer when they readily put out roots and will have time to establish a good root system before winter. The cut end should be dusted with aluminium powder or a hormone rooting powder, or exposed to direct sunlight for several hours. This promotes more rapid healing of the cut surface and helps prevent infection. The shoots should be sufficiently large (at least the size of a hazel-nut) so that they will not dry out before producing roots. When the cut dries, after about a week, place the shoot on the surface of a peat and sand mixture and keep the mixture slightly moist. The shoot may also be laid in a dry place and put in the soil after it begins to form roots. In both instances, the shoot or stem cutting should be placed in a warm, shaded spot with higher relative humidity. Once a new root system has formed, water the plant frequently (two to three times a week) but with great care. The newly formed roots are very delicate and if the substrate is excessively wet and there is a sudden drop in temperature, they readily rot.

Propagation from seed

Propagation of cacti from seed is the commonest means of reproduction in the wild and thus also a common method in cultivation. Propagation from seed yields a greater number of

18

seedlings that are adapted to the given environment, and allows the grower to choose healthy and typical specimens for further cultivation. Germinating plants and small seedlings must not only be provided with heat, light and moisture, but also with a sufficiently long time in which to grow so they are able to attain sizeable dimensions before the onset of winter, thereby enabling them to survive this period without harm. If the seeds can be placed indoors near a window that faces south, south east or south west, then they may be sown in late winter or early spring. If no such position is available in the grower's home, he or she should wait until May and sow the seeds in a glass case or else in a special glass propagator illuminated and heated by a fluorescent lamp. The temperature required for the germination and a good growth of the seedlings is 25—30°C (11—86°F) with a possible slight drop during the night. Recommended for this purpose are plastic boxes (seed trays) at least 5 cm (2 in) deep, furnished with a transparent cover. On the bottom of the box place a drainage layer of washed coarse sand or fine stone rubble (not limestone), about 1 cm ($\frac{1}{3}$ in) thick, which will later make it possible to water the substrate from below. Top this with the growing medium, which may consist of equal parts of soil, sand and peat, or else equal parts of soil, sand, peat and crushed brick, and also a mixture of one part soil and one part river sand. The proportion of the various ingredients is not so desperately important; what is important, however, is that the substrate is loose, will not become compacted and does not contain decomposing organic substances that might later promote the growth of moulds that would pose a great danger to the growing seedlings.

All the ingredients should be thoroughly mixed and the substrate disinfected or, best of all, sterilized with steam. To do this, take an old, discarded cooking pot, pour in a little water, spread a layer of sand over the bottom, put in the prepared growing medium and heat it to about 100°C (212°F) for at least one hour. When the growing medium has cooled, spread it evenly over the drainage layer in the box and press it down lightly. It should be about 2.5 cm (1 in) thick. The final step is to take previously cut strips of plastic sheet and divide the box into separate sections.

This done, you can proceed with the sowing. Put the seeds in a test tube, add a minute quantity of some fungicidal agent, cover the opening of the test tube with your thumb and shake the contents vigorously. Then sow the seeds evenly on the surface of the growing medium. Cactus seeds should never be covered with soil; only lightly press in the largest seeds. When all the seeds are sown, let the substrate soak up water from below. You may add a bit of fertilizer to the water so that the germinating plants will have sufficient nutrients from the very beginning, especially in poorer substrates.

If you have a large number of seeds, you can sow them in small square flowerpots. On the bottom of the flowerpot put a layer of drainage and then fill it up with the growing medium. After sowing the seeds in the rooting pots, place these in a box with a transparent cover. Water from below by taking out one of the rooting pots and pouring water into the empty space. One great advantage of sowing in rooting pots is that it allows for the separate handling of each species.

Cover the box with a transparent lid and put it in a warm, light place, shaded from direct sunlight. One of the great dangers with this set-up is that the inside of the box could easily become overheated. Throughout the entire germinating period, as well as during the seedlings' first few months, the soil must never be allowed to dry out. Ventilate the box only if the plants show signs of fungus rot. Remove damaged seedlings with forceps, together with the soil around the spot where they grew, and sprinkle the area with a fungicidal solution.

The recommended time for the first transplanting or pricking out of the seedlings is after three to six months, when they have attained such dimensions that they are crowding each other, and have developed a skin that is thick enough to withstand the damage of squeezing with forceps. The substrate to which they are moved should also have been sterilized with steam and should be slightly moist. If the seedlings already have a relatively developed root system, leave them in the open for at least two days after removal and prior to replanting so that any damaged or shortened rootlets can dry. Proper spacing of the seedlings is important — they grow best if the distance between them equals the diameter of their stems.

Grafting
This method of propagating cacti is very common and is an integral part of modern cactus cultivation. In grafting, a suitable portion of a given plant (scion) is joined to a portion of another plant (stock). The two parts then become united and form a single plant. Grafting is used for several reasons:

1 To speed up the growth of slow-growing species.
2 To facilitate the cultivation of cacti with a delicate root system.
3 To produce earlier and more profuse flowering.
4 To save plants whose roots or the lower parts of the stem have been attacked by rot.
5 To propagate cristate or other unusual forms as well as species that are difficult to propagate from seed.
6 To cultivate mutants without chlorophyll.

Rootstock
For aesthetic reasons, the preferred choice is low stock; the only exception is in the case of stock for specimens with downward-hanging stems (e. g. *Epiphyllum, Rhipsalis* and the like), for specimens with stems that must not come in contact with the soil (e. g. cristate forms), or for grafted cacti that are dependent on the stock for their supply of food (coloured forms without chlorophyll). Sometimes the stock is gradually exhausted by the scion, which then roots on its own. At other times it is possible to cut the scion off once it has attained the desired dimensions and then let it root on its own.

The rootstock that is generally used and that is suitable for practically all species of cacti, especially for Mexican rarities, is *Echinopsis eyriesii*. This undemanding cactus has proved to be good for grafting small seedlings as well as adult specimens. It will grow readily, even in winter, and is therefore suitable for 'rescue' grafting. Another commonly used and hardy stock is *Eriocereus jusbertii*. It, too, is suitable for most species, which will thrive on it exceedingly well. The only problem is that grafts on *E. jusbertii* do not take as well, particularly when it is in the most active stage of growth. Recommended as permanent stock for many, and especially South American, species is *Cereus peruvianus*, which, however, must be in the full stage of growth at the time of grafting. Another very good stock is *Pereskiopsis*, on which the growth of scions is more rapid than on any other stock. Its one drawback is that it requires a light, relatively moist and warm environment during the winter season.

Methods of grafting
The best time for grafting is during fine weather in spring. The commonest method of grafting cacti is when the stock and scion are cut at right angles to the axis. First, prepare the stock. Remove the top of the stem by making a smooth horizontal cut

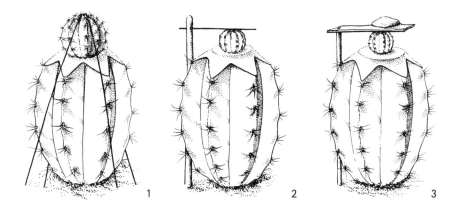

Various methods of fastening a scion (graft) to the stock:
1 by means of a rubber band; 2 by means of a flexible wire;
3 by means of a weight

with a sharp, stainless steel knife about 1 cm below the crown. Then trim the edges of the ribs at a slant round the periphery into the shape of a truncated cone. The first cut is rarely straight and perfectly smooth, therefore repair the surface by making a second cut. Should the second cut also be unsatisfactory, make yet another cut. The cut surface and main vascular bundle of the stock must be healthy. If there is even one reddish spot in the vascular bundle, the stock should be discarded. Replace the removed piece of stock on the cut surface to prevent it drying out and set about preparing the scion. The procedure is the same. Cut the scion at right angles to the axis, trim the edges of the cut surface, and if necessary make a second cut. The cut surface of the scion must likewise be straight and smooth, and must be made with a clean knife.

Next remove the piece covering the cut surface of the stock and place the scion on it. Ideally the stock and scion should be of the same diameter, but this is not essential. The important thing is that the vascular bundles (each in the form of a circle in the centre of the cut surface) coincide with one another as closely as possible. If these circles are not the same size, place the scion so that the edges of the vascular bundles meet at least on one side or else intersect. Then turn the scion a little, exerting slight pressure with the hand to force air out, and fasten it securely to the stock by means of rubber bands. This requires a practised hand so that the scion does not slide off or get pushed off by the rubber band. Newly grafted plants should be left in the sun for about an hour and then placed in a shaded but sufficiently warm and dry place. The rubber bands may be removed after about a week.

DISEASES AND THEIR PREVENTION

Cacti that receive proper care are much more resistant to disease and attack by pests than cacti that are neglected, not repotted regularly, or are placed in an unsuitable position. The most widespread diseases are physiological diseases or, rather, physiological disorders. Every flouting of the basic rules for the cultivation of cacti, every mistake in the growing conditions, affects their growth. The following are the commonest causes of physiological disorders.

1 **Intensive exposure to the sun** Intensive exposure to the sun and high temperatures lead to sunscorch, especially if the plant is in an enclosed space. The skin of the cactus begins to turn pale or redden, and after a few hours will even turn white and cave in on the overexposed side. Plants most readily affected by overexposure to the sun are those that have not yet become accustomed to strong sunlight, this being in spring and particularly in the case of specimens overwintered in the dark. Also readily affected are cacti not yet in their full stage of growth, whose tissues are not sufficiently saturated with water. The best care is prevention. In spring, shade the plants, for example by whitewashing the glass. In summer, provide them with adequate ventilation, and in the case of cacti that are in the full stage of growth (e. g. following the loss of roots), move them to a shadier spot or cover them with paper.

2 **Too low a temperature** A drop in temperature during the growing season halts the growth of cacti and, if the soil is excessively moist, causes rotting of the roots. Cacti should be watered only during warmer periods. During overwintering the temperature should not drop permanently below 5°C (41°F) and for thermophilous species (*Melocactus, Discocactus, Uebelmannia, Pilosocereus*, etc.) it should be no less than 12—15°C (54—59°F). A lengthy period of temperatures below these limits, accompanied by higher atmospheric moisture, generally leads to fungal diseases. Brief night-time

drops in temperature, even below freezing point, during spring and autumn, however, are not harmful as a rule.

3 **Overwatering**   Permanent overwatering of the substrate causes decay of the roots and sometimes also of the base of the stem. Here, too, prevention is a must — see the points discussed in the sections 'Planting and Repotting' (page 13) and 'Watering' (page 15). If the roots do decay, take the cactus out of the pot immediately, wash the roots or remove the soil particles by hand, and cut back the roots to healthy tissue, without any red veins in the centre of the root. It is often necessary to divest the cactus of all its roots and let it put out new ones. If there are signs of rot on the root neck, you must cut out the damaged tissue, let the wound dry in the sun and possibly also dust it with a fungicidal preparation. Then repot the plant at a shallow depth, leaving the root neck exposed. In the case of more extensive damage, remove the entire root neck and let the plant root again by itself or graft it on stock.

4 **Atmospheric moisture**   On the one hand, atmospheric moisture promotes more rapid growth of cacti; on the other, it contributes in large part to the spread of fungal diseases. By airing glasshouses and frames when temperatures are at their peak during the day, as well as at night, you will curb fungal attack and will make the cacti hardier and more resistant to disease. Airing will also help to prevent fleshy fruits and seeds from being attacked by mould, which often happens in the autumn. Not enough atmospheric moisture, however, is also undesirable: when the air is very dry, plants should be misted.

5 **Lack of light**   If they do not have enough light during the growing period, cacti grow tall, have short, sparse, indistinctly coloured spines and do not flower. They must be moved to a more suitable position with a minimum of four to six hours of sunlight during the daytime. During the resting period, however, cacti may be kept in complete darkness; deformations will be prevented at this time by providing absolutely dry conditions and a temperature of around 5—15°C (41—59°F).

6 **Chemical reaction of the soil**   Excessive acidity or excessive alkalinity of the soil causes a slowdown or the complete cessation of growth. The soil should be slightly acid or neutral with a pH of approximately 5.5—7.0.

Another important group of diseases are those caused by fungi.

24

These are known as mycoses. The most widespread disease of small seedlings is damping-off (also called propagator disease), which is caused by a fungus that first forms a coating on the surface of the soil and later penetrates the tissues of germinating plants, as well as older specimens, causing them to disintegrate. It attacks cacti immediately beneath the surface of the soil in the region of the root neck. The affected plants flop to the side on which they are invaded. Damping-off generally affects germinating plants or seedlings grown in the same box. When this happens, air the interior of the box, remove the affected seedlings and spray the soil from which they have been lifted with a fungicidal preparation. In older specimens remove the diseased tissue with a knife, dust the wound with a fungicide and, after several days, put the plant in new soil, leaving the root neck exposed.

Another fungal disease to which cacti are commonly prone is root rot, caused by *Phytophthora omnivora* and other fungi. This is characterized at first by the decay of the roots and base of the stem; later by decay of the whole body. In such a case, lift the plant from the soil immediately, cut off the diseased roots and parts of the stem to healthy tissue, and let the wound dry in the sun.

PESTS

The commonest pests of cacti are coccids, mainly the root mealy-bug (*Rispersia falcifera*), which is parasitic on the roots. It moves about freely over the roots before forming a protective cotton-like covering of white waxy fibres inside which it reproduces. It may produce as many as seven to ten generations in a single growing season but the low temperatures of the winter months bring about a partial or complete cessation of the insects' development. Root mealy-bug not only damages the roots by sucking the sap, but may also cause the decay of cacti if the stems are attacked by a fungus via the damaged roots. This is a very common pest and one that is extremely difficult to eradicate in large collections. When you bring home a new specimen, always remove it from the pot and examine it thoroughly to check for signs of root mealy-bug. If you discover white wax-like wads on the roots, immediately cut them off, together with the terminal rootlets, and remove them from the axils of the main roots. Then immerse the whole plant in an insecticidal organophosphate solution. Use the same solution when you first water the plant after putting it in soil. Recommended as a preventive measure is spraying the whole collection with a proprietary insecticide every fourteen days in the autumn or spring.

Planococcus citri and *P. adonidum* are the coccids most commonly parasitic on the stems of cacti. Their soft oval bodies are protected by waxy wads resembling wool and that is why they are called wooly aphids by cactus growers. They generally occur in sheltered places on the body — between the ribs, between mamillae, on the underside of areoles, etc. The larvae are active and climb onto other host plants on which they feed during their development. When the temperature drops in winter, they descend to the base of the plant and onto the main roots. Here, too, the best method of control is spraying with an insecticidal organophosphate solution. However, when a plant is greatly infested it is best to submerge it totally in the respective solution. Another common parasite is the red spider mite (*Tetranychus urticae*), a minute insect less than 0.2 mm long. It moves about rapidly over the stems of cacti, piercing the skin and sucking the sap. It also disfigures the plant with pale spots that later turn rusty and expose the plant to the danger of secondary viral, bacterial or fungal infection. Red spider mites are also controlled by misting with an insecticidal organophosphate solution.

Organophosphate preparations are poisonous substances, harmful to health, and must be handled accordingly. Always wear rubber gloves and provide good ventilation of an enclosed space. Spraying with these preparations is harmful to cacti if done during a hot period and in sunny weather. Last but not least, it is recommended not to use only one preparation but to alternate this with other preparations so as to prevent the development of resistant strains. Spray the plants a second time after an interval of ten to fourteen days in order to destroy any pests that survived the first application in the egg stage.

26

COLOUR PLATES

CACTI OF NORTH AMERICA

Aporocactus flagelliformis (L.) Lem. — Rat's Tail Cactus

A. flagelliformis has been cultivated as a houseplant in Europe since 1690. Linnaeus gave it the name *flagelliformis*, which means whip-like, in reference to its long slender stems. It is an epiphyte, found growing in a number of places on the highland plateaux of Mexico. The cactus forms bushes consisting of trailing stems 1—2 cm ($\frac{1}{3} - \frac{3}{4}$ in) thick and up to 1.5 m (5 ft) long. The spines, ten to fifteen in each areole, are golden-yellow, reddish-brown or brown. The zygomorphic flowers are dark pink, up to 8 cm (3 in) long, and are pollinated in the wild by hummingbirds. In cultivation, they bloom in spring, remaining open for about four days. A number of species of *Aporocactus* have been described to date besides the one illustrated here, but these differ only in quantitative aspects and so it would be more suitable to class them as varieties or forms. These are undemanding plants noted for their rapid growth and profuse flowering. Nevertheless, they have certain specific requirements as regards cultivation. They do best in a shaded position in a glasshouse or glass-fronted veranda, by the window among other houseplants, or in the garden in summer. Like other epiphytes, *A. flagelliformis* requires a loose, well-aerated substrate composed of equal parts of rich soil, sand and peat, and cooler conditions in winter (6—15°C/43—59°F), along with greatly limited watering. In winter, keep the plant in a light place as it will begin to put forth flowers early in the spring.

2

Aporocactus flagelliformis (1) attracts notice mainly during the flowering period when thicker clumps may bear a vast profusion of blooms. For this reason it was often cross-bred to obtain even larger and more colourful flowers. Examples are the *Aporophyllum* hybrids developed by crossing with *Epiphyllum*, and the beautifully flowering *Helioaporus*, a hybrid obtained by crossing *A. flagelliformis* and *Heliocereus speciosus* (2). *H. speciosus* grows in rocky terrain near Mexico City, its stems, up

28

1

to 1 m (3ft) long, trailing over boulders
or hanging from rock outcrops.
Sometimes it grows as an epiphyte in
treetops. Its flowers are up to 15 cm
(6 in) long, wide-spreading, and
coloured carmine-red with
a violet-opalescent sheen.

Astrophytum capricorne (A. DIETR.) BRITT. ET ROSE

The genus *Astrophytum* comprises five species that are readily distinguished from each other, plus numerous varieties. Characteristic features are the white flakes thickly covering the skin, and yellow flowers.

A. capricorne, like the other closely related species, grows on rocky limestone substrates and gravel waste in the north Mexican state of Coahuila. In collections it is often incorrectly labelled *A. capricorne* var. *maior*. It is a globe-shaped, eight-ribbed cactus up to 30 cm (12 in) high and 15 cm (6 in) across, thickly covered with flakes. An important character of identification is the rust-coloured, newly formed flakes in the area of the vegetative centre. The flowers are golden-yellow with a red centre and are a typical characteristic of other astrophytums native to Coahuila as well. Chief of these is *A. capricorne* var. *minor*, which is smaller, has white flakes in the area of the vegetative centre, and softer, more numerous spines that are often a pale colour.

Rarely occurring specimens bearing flowers without a red centre are classified as *A. capricorne* var. *minor* cv. *crassispinoides*. Another variety is var. *niveum*, previously classed as a separate species, *Astrophytum niveum*. This plant was originally named in reference to the thick cover of flakes but specimens were later found that were only partly covered with flakes or were even without any flakes whatsoever.

A. capricorne var. *niveum* attains a height of 0.5 m ($1\frac{1}{2}$ ft) and has thick, firm spines. *A. capricorne* and its varieties are relatively prone to damping-off and for that reason coarse sand or gravel should be spread over the area round the root necks of seedlings as well as adult plants. They require a well-draining, mineral substrate and need watering only in very warm and constant weather.

Astrophytum capricorne (1) has about five spines growing from each areole; these are flat, up to 3 mm wide, and coloured dark brown, later turning grey. They begin to appear on the plant about the third year; until then the seedlings are without spines. Spineless specimens (2) may occasionally be encountered even among adult plants. *A. capricorne* var. *senile* is a striking variety, distinguished

3

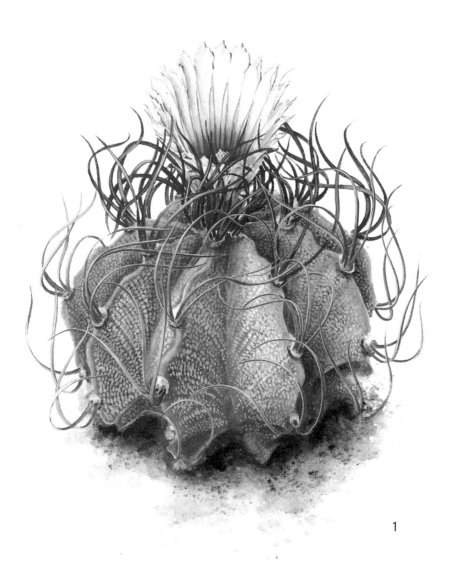

1

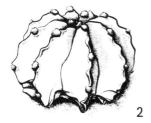

2

by having a nude skin devoid of flakes
or with only very few flakes and about
twenty spines growing from each areole.
The spines are up to 10 cm (4 in) long
and greatly twisted and intertwined.
They are already present on small
seedlings and are dark brown or yellow
when they emerge. Plants with yellow
spines are named *A. capricorne*
var. *senile* f. *aureum* (3).

Astrophytum myriostigma LEM. — Bishop's Cap

Few cacti have a common name in addition to their Latin designation. *A. myriostigma* is known as the Bishop's Cap among cactus growers all over the world. It was introduced to other countries from Mexico where it is relatively abundant on calcareous soils in the states of San Luis Potosí, Tamaulipas and Nuevo León, although it has already become extinct in some places. Plants native to San Luis Potosí may be designated as the type variety; these were originally named *A. myriostigma* var. *potosínum*. However, the correct name is *A. myriostigma* var. *myriostigma*. They have a flat-topped globose body up to 30 cm (12 in) across, covered with white-felted spots and with sharp-edged or rounded ribs without spines. Individuals with rounded ribs were later described as var. *strongylogonum*, but because they grow together with sharply ribbed specimens in the wild and their offspring do not have stable, hereditary constant characteristics, they cannot justifiably be classed as a variety.

A. myriostigma requires a well-draining substrate, plenty of heat and light in summer and overwintering at a temperature of about 10°C (50°F) in an absolutely dry substrate. Watering should not be started until late spring when the glasshouse temperature is at least 30°C (86°F) or higher.

2

Astrophytum myriostigma (1) has several varieties. Var. *myriostigma* retains its broadly globose shape even in older specimens and of all the varieties it has the largest flowers, measuring 5—8 cm (2—3 in) across when fully open. Typical of all the species are the white-felted flakes on the skin, although some populations of *A. myriostigma* were found to be only very sparsely covered with these flakes or to lack them altogether. Selected nude specimens were then repeatedly cross-pollinated until the process yielded offspring in which this characteristic was constant. These are designated *A. myriostigma* var. *nudum* (2). A striking

1

variety is var. *tulense* (3) from the
neighbourhood of Tula, a town in
Tamaulipas state, after which it is often,
invalidly, named var. *tamaulipense*. It
differs from the type variety by its
columnar habit and the arrangement of
the areoles, which are close together and
in older specimens sometimes even
merge to form a continuous felt border.
Mention should also be made of var.
columnare, which has a narrowly
columnar stem from the earliest stage of
growth, and also var. *quadricostatum*, at
a later date (and hence invalidly) named
var *jaumavense* after the place of its
occurrence. This variety generally has
only four, very occasionally five ribs.

3

Astrophytum ornatum (DC.) F.A.C. Weber

A. ornatum is the longest-known (since 1827) and also the most southerly-occurring species of this genus. It is native to the Mexican states of Hidalgo and Querétaro, where it grows on rocky slopes and very occasionally also on flat land. It attains a height of 1.5 m (5 ft) and a diameter of 30 cm (12 in) and has eight ribs, which is a very conservative number. However, it exhibits marked differences in the arrangement and density of the flakes. The flakiness often changes during the plant's development — young seedlings are usually more thickly spotted, adult specimens may even be entirely nude. According to some authorities, this flakiness is of particular importance in the case of young plants, as the flakes are capable of absorbing atmospheric moisture, thereby supplementing the supply of water to the plant tissues at a time when this function cannot be satisfactorily fulfilled by the newly developing root system. The flakes also serve as a reflective layer that protects the plant against the intense rays of the sun. The spines of *A. ornatum* are quite variable. In some plants they are golden-yellow, in others yellow-brown or dark brown. They are usually about 4—5 cm ($1\frac{1}{2}$—2 in) long, but sometimes even more than 10 cm (4 in). The flowers that are produced on older plants are pure yellow and when fully open may be up to 9 cm ($3\frac{1}{2}$ in) in diameter. As in other astrophytums, they open during the forenoon over a period of two to three days. The seeds of all astrophytums are large and germinate well, usually two to three days after sowing. Because the seedlings do not tolerate undue moisture the space where they are grown should be aired early on. For that reason it is recommended to sow the seeds separately, keeping them apart from other later-germinating species.

2

Several varieties of *Astrophytum ornatum* (1) have been described to date on the basis of differences in the flakiness of the skin. All, however, are unjustified from the botanical viewpoint for they occur intermixed throughout the whole range of the species' distribution. In some specimens the spots are distributed regularly over the skin, in others they

1

3

are arranged in bands of varying width (2) or else are absent altogether (3). Cross-pollination of these nude individuals in cultivation gave rise to the cultivar designated cv. *virens*, i.e. green. At the other extreme is the cultivar *niveum*, i.e. snow-white, so thickly spotted that the spots merge into a solid cover.

Cephalocereus senilis (HAW.) PFEIFF. — Old Man Cactus

C. senilis is a classic representative of white-haired cacti. This apt name is similar to the Mexican name 'cabeza del viejo', meaning old man's head. It grows in Hidalgo state on the steep slopes of the deep valleys of the Tulancingo River system, where the climate is warmer than on the surrounding Mexican plateau into which these valleys are carved. Nowadays its numbers are greatly depleted because for over 150 years small specimens capable of being transported were widely collected and other giant specimens were felled for their seeds. Currently *C. senilis* is protected by the Mexican government and thus there is hope that it will be preserved for future generations. *C. senilis* is a typical columnar cactus that does not produce offshoots; it reaches a height of 10—12 m (33—39 ft) and measures 18—24 cm (7—9 in) in diameter. It attains sexual maturity on reaching a height of about 6 m (19½ ft), when it begins to form a lateral cephalium from which pale pink flowers, 7.5 cm (3 in) across, emerge. Its cultivation is moderately difficult. In winter it requires dry conditions and a temperature of about 10—15°C (50—59°F); in its native land it grows only up to elevations of 1,600 m (5,250 ft) and is not found in the harsher climatic conditions of the Mexican highland plateau. In summer it should be placed in a sunny position. It should be provided with a free-draining mineral substrate and watered only in higher temperatures when the substrate is thoroughly dry.

2

Cephalocereus senilis (1) is protected against the intensity of the sun's rays by a thick covering of 10- to 20-cm-long (4—8 in) hair-like spines. Mexico is also the home of other, very ornamental 'white' cereus cacti. These, however, do not have a protective covering of white hair but only a light-coloured skin. Easy to grow, for example, is *Myrtillocactus geometrizans* (2), native to southern and central Mexico. Adult plants form much-branched trees measuring 3—4 m (10—13 ft) in width as well as height. The individual branches measure 6—10 cm (2⅓—4 in) in diameter and are covered with a lovely bluish bloom at

1

3

first. The numerous greenish-white flowers (3), produced near the top of the stem, are 2.5—3.5 cm ($1-1\frac{1}{3}$ in) wide. The fleshy fruits are coloured bluish-red, hence the vernacular name 'cranberry cactus'. In cultivation this species is also used as stock for grafting, especially for coloured cultivars without chlorophyll.

Coryphantha calipensis H. Bravo

The name *Coryphantha* means flowering from the top. This is not the only characteristic coryphanthas have in common. Another typical feature is the division of the stem into prominent mammillae, which have a noticeable groove on the upperside along their entire length, from the areole to the axil. The flowers of coryphanthas always grow on the crown from these grooves.

One of the most interesting species in appearance is that illustrated, *C. calipensis* from the state of Puebla in Mexico, where it grows in the neighbourhood of the town of Calipán. The stem is globose at first, and later, in adult plants, slightly elongated with its conical crown thickly covered with wool. The skin is greyish to olive-green. The narrow, upward-directed mammillae overlap like the scales of a pine cone, giving the plant a distinctive aspect. The flowers are approximately 6 cm ($2\frac{1}{3}$ in) across, and coloured lemon-yellow with red filaments. They are produced repeatedly throughout the summer, remaining open for about four days. As in all coryphanthas, they are allogamous. *C. calipensis* is susceptible to sun scorch and thus should be placed in a warm but partly shaded spot. To preserve the plant's rich wool — one of its greatest ornaments — it should be watered from below. In winter it requires cool and dry conditions. Propagation is by means of seeds.

3

Coryphantha calipensis (1) is an interesting, unmistakable plant only in adulthood, when its stem resembles a giant cone. In small seedlings (2) the mammillae are of an entirely different shape so that they do not resemble adult specimens. Most closely related to the illustrated species is *Coryphantha cornifera* (3), likewise from central Mexico, which has a globose stem up to 12 cm ($4\frac{3}{4}$ in) high. Typical of this cactus are the large, dark brown to black, horn-shaped, downcurved central spines, from which the plant takes its name (*cornifera* means horned). The fleshy fruits of coryphanthas, coloured green when ripe (4), are produced the year

1

2

4

following fertilization. The fruits should be picked only after they have loosened their hold and can be pulled off with ease, not forcibly. Shortly afterwards, squeeze out the seeds and wash them; do not wait too long, for the seeds are hard to remove when the fruits have become dry and hard.

Coryphantha elephantidens (Lem.) Lem.

Several related species of *Coryphantha* grow on the fertile grasslands around Mexico City. They are very similar and have the same requirements in cultivation. Typical characteristics are robust mammillae, larger dimensions and yellow flowers. The only exception is *C. elephantidens*, which has pale to deep pink flowers. Adult plants attain a width of approximately 20 cm (8 in) and a height of 15 cm (6 in), which makes this cactus the largest of its genus. It produces offshoots only rarely and definitely does not have the tendency to form groups. The spines are quite variable, in length as well as width. With their shape and greyish-white colour, they resemble elephant tusks and that feature is reflected in the plant's specific name. Another typical characteristic is the white wool filling the hollows between the mammillae, which grows most abundantly and has its greatest staying power in the area of the vegetative centre at the top of flower-bearing specimens. The flowers emerge from this wool on the crown during the latter half of the summer. Compared with the blooms of other coryphanthas, these are the largest, measuring about 10 cm (4 in) in diameter. Not all coryphanthas are easy to grow in cultivation, but this does not apply to *C. elephantidens* and closely related species. In their native land these species are used to humus-rich, nourishing soil and large quantities of water during sudden downpours. If these conditions are provided in cultivation, their growth is rapid, without much danger of them dying, and they will begin to bear flowers in the fourth or fifth year. Best suited is a warm, only slightly shaded position, and a winter temperature of 5—15°C (41—59°F).

2

The flowers of *Coryphantha elephantidens* (1) are quite variable in coloration. Usually they are pale pink with a red throat, red filaments and red base of the pistil. In some specimens, however, the perianth segments may be almost white or, conversely, a delicate to deep red. Closely related species include *C. sulcolanata, C. andreae, C. greenwoodii* and above all *C. bumamma*. The stem of

1

C. bumamma in youth (2) is practically bare, in adult plants it is conspicuously covered with wool. This, however, is readily washed off when the plant is watered from above. C. bumamma differs from C. elephantidens in the size of the stem, which is up to 28 cm (11 in) in diameter, and also in its flowers, which are smaller and always yellow. Its spines are very variable.

Echinocactus grusonii Hildm. — Golden Barrel Cactus

E. grusonii is the most familiar and most ornamental of the globose cacti. Since it was discovered in Mexico in 1889, countless specimens of various sizes have been exported, mostly to the southern USA where these 'golden barrels' were ideal garden ornaments. In the 1970s, however, their export was greatly limited. Nowadays *E. grusonii* is only sparsely distributed in the Rio Moctezuma river system. This river forms 500- to 1,000-m-deep (1,600—3,300 ft) canyons with a more humid and warmer climate that is quite different from the harsher climate of the surrounding plateaux of Querétaro and Hidalgo states. Sadly, the last investigations showed that the population density of this species has currently dropped below the critical limit so that as things stand the plant is on the way to becoming extinct. Besides collecting, this state of affairs has been brought about by insect pests, mainly snout-beetles whose larvae feed on the fruits and seeds. However, inasmuch as there are already many flowering specimens of *E. grusonii* in cultivation, the production of seeds for commercial purposes is assured. The bright yellow stem of *E. grusonii* is up to 80 cm ($31\frac{1}{2}$ in) high and 70 cm ($27\frac{1}{2}$ in) across and has numerous longitudinal ribs. Each areole has about fourteen spines, 4—6 cm ($1\frac{1}{2} - 2\frac{1}{3}$ in) long and transversely grooved (annulate). This cactus is relatively easy to cultivate. During the growing period it requires adequate application of water, and in winter dry conditions and a temperature above 10° C (50° F). In spring, when it is more susceptible to sun scorch, it should be well shaded.

2

Echinocactus grusonii (1) does not flower until an advanced age, when it measures at least 50 cm ($19\frac{1}{2}$ in) in diameter. The flowers are therefore seldom produced in central Europe. The yellow, autogamous flowers, which are not particularly striking, grow from the youngest areoles near the crown. However, this cactus is decorative even without flowers. Its chief ornament is the stout spines, translucently golden-yellow to pale yellow with a darker base. In collections one will

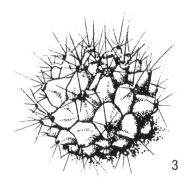

also come across mutational forms with white spines (2) called f. *alba*, or with reduced spines. The small seedlings (3) do not show much of a resemblance to adult specimens, for the stem is composed of tubercles that merge into ribs only as the plant gets larger, at the age of about four years.

3

1

43

Echinocereus delaetii (GÜRKE) GÜRKE

The genus *Echinocereus* includes some 40 species and numerous varieties, many of which were at one time considered to be independent species. They are plants of short columnar habit, some of which produce offshoots and form huge clumps. The flowers are large and coloured reddish-violet, or very occasionally yellow.

E. delaetii is found in Mexico in the southern part of Coahuila state, where it grows in calcareous soils at elevations of about 2,000 m (6,500 ft). With its long 'hair' it differs markedly from the other members of the genus and for that reason was originally described as *Cephalocereus delaetii*. Its resemblance to the seedlings of *C. senilis* is truly remarkable. This, however, is an example of mere convergence, i.e. similarity, not kinship, which the author of the description soon realized; he then re-classed the species in the genus *Echinocereus*. *E. delaetii* produces numerous offshoots, forming clumps composed of up to 50 erect or prostrate stems. The stems are approximately 30 cm (12 in) long and 4—8 cm ($1\frac{1}{2}$ — 3 in) across. The plant's most noteworthy feature is the hair-like, 6- to 10-cm-long ($2\frac{1}{3}$ — 4 in), curving spines. As in most members of the genus, the flowers are large, measuring about 6 cm ($2\frac{1}{3}$ in) in length as well as width, and coloured pale to purplish-pink. The plant flowers very poorly and *E. delaetii* is propagated by means of offshoots as well as seeds. If you decide to grow plants on their own roots (i.e. not grafted on stock), which is a less common method, put them in a sandy, free-draining substrate and apply water only when this is thoroughly dry. Like most rather delicate echinocerei, this cactus requires a summer resting period. It likes a sunny, warm place and in winter cooler conditions with a temperature of 5—12°C (41—54°F).

3

The stem of *Echinocereus delaetii* (1) is covered with white spines that are densely intertwined in older specimens, a characteristic that makes identification easy, even by the inexperienced cactus grower. Other white-spined echinocerei do not have such trailing 'hair'. Of the closely related species, mention should be made of *E. freudenbergeri*, sometimes recognized only as a variety of *E. delaetii*, and *E. longisetus* (2), also

from northern Mexico. The latter,
however, differs from *E. delaetii* by
having straight, bristle-like spines, as
indicated by its name. Another beautiful
species is *E. nivosus* (3), also often
known by the invalid name *E. albatus.*
In the wild it forms clumps composed of
numerous short stems covered with
glassy white spines. Its flowers are small
and red.

Echinocereus knippelianus LIEBN.

E. knippelianus, with its peculiar spines and broad, flat ribs, is one of the most interesting cacti. It is native to Mexico, to Coahuila state, where it grows on open grasslands and at the edges of pine woods at elevations of approximately 2,000 m (6,500 ft). Whereas it grows singly in the wild, only very occasionally forming clumps, in collections it is often the other way around. This habit in cultivation is caused by the better conditions — more water and more food — generally made possible by the stock on which it is grafted. The stems measure up to 8 cm (3 in) across and whereas in its native land they rise only slightly above the level of the surrounding terrain, in collections they reach a height of approximately 10 cm (4 in). Characteristic of this species are the stout turnip-like root, dark green colour of the skin and broad, flat ribs. The spines, only one to three in each areole, are yellow and about 1.5 cm ($\frac{1}{2}$ in) long. The flowers are pale pink to pink, very occasionally nearly white, and approximately 3 cm (1 in) long. They emerge from lateral areoles in early spring. *E. knippelianus* loses its roots if watered too much and in an inappropriate manner and these only grow in anew with great difficulty, hence it is recommended that it be grafted. It requires a heavier, sandy-loamy substrate without a greater proportion of organic particles and more liberal application of water only in spring and late summer. It should be placed in a spot with sufficient sunlight and wintered at a temperature of about 10°C (50°F).

3

Echinocereus knippelianus (1) is beautiful and distinctively different from other cacti even without flowers. Besides the type variety *E. knippelianus* var. *knippelianus*, there are two others, described at a later date. One is var. *kruegeri* (2) with three to four slender spines, about 5 cm (2 in) long, to each areole and with flowers that are generally whitish and are produced on the crown. The second is var. *reyesii* (3), which has four stiff, prickly, needle-like spines to each areole. The flowers likewise emerge on the crown but, unlike those of var. *knippelianus* and *kruegeri*, measure up to 6.5 cm (2$\frac{1}{2}$ in)

1

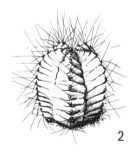

2

across and are coloured a deep pink. The plant flowers readily and quite soon, generally as early as the fourth year. It is native to Nuevo León state, where it grows at elevations of approximately 3,000 m (9,800 ft) and, like most mountain species, grows reliably on its own roots.

Echinocereus pectinatus (S<small>CHEIDW</small>.) E<small>NGELM</small>.

E. pectinatus is the most familiar of the so-called pectinate echinocerei. It derives its name from the pectinate (i.e. comb-like) pattern of its spines. It occurs in a number of forms and varieties throughout a large area in nothern Mexico and southwestern USA, where it grows in semi-deserts and savannas at elevations of 800—1,900 m (2,600—6,200 ft). It generally occurs singly, only very occasionally producing offshoots. The stem is 8—15 cm (3—6 in) high. The flowers are up to 7 cm (2¾ in) across and coloured pink with a paler centre. Like most related species, it is not easy to cultivate and therefore it is recommended that it be grafted on low stock — *Echinopsis, Trichocereus, Helianthocereus, Eriocereus,* and the like. Grafted plants grow rapidly and flower early, and the danger of them dying is reduced to the minimum. Problems occur at a later stage of growth, however, when the plants become too long and prostrate and they should then be rejuvenated by regrafting the top parts. Excessive prolongation of the stem may be partially forestalled by limiting watering during the growing period. If you want to grow *E. pectinatus* on its own roots, put it in a mineral substrate and apply water only when this is thoroughly dry. Do not water the plant when the temperature drops or during the hottest period in summer, when *E. pectinatus* has a second resting period. In summer, it requires a sunny and warm location; in winter, a temperature of about 10°C (50°F).

3

The stem of *Echinocereus pectinatus* (1) is completely covered with close-fitting pectinate spines.The spines (2) are dense and during the resting period, when the stem dries, provide excellent protection against sun scorch. The flowers remain slightly closed even when in full bloom. As in all members of the genus, the flower tube and receptacle are covered with bristles and spines. *E. pectinatus* occurs in several varieties, of which var. *rigidissimus* (3) is the most familiar. Its spines are not only rigid but also variously coloured at different levels of the stem, which gives it a striped appearance — hence its common name of 'rainbow cactus'.

1

2

49

Echinocereus reichenbachii (TERSCH. ET HAAGE)

Some cacti can be grown outdoors even in Central Europe if they are provided with a south-facing location with good drainage. Grown thus are not only the hardiest opuntias but also other cacti — mainly several species of *Escobaria, Pediocactus* and *Echinocereus.* One such hardy echinocereus, which tolerates a winter temperature drop even below −20°C (0°F) if provided with dry conditions, is *E. reichenbachii.* This cactus is found growing from practically sea level to elevations of 1,500 m (5,000 ft) in north-eastern Mexico as well as in several states in the south-western USA, in other words in regions where summers are extremely hot and winters extremely cold but dry. It is distributed throughout a very wide range and this is reflected in its variability. For this reason some of the most distinctive geographical deviations were once described as independent species, such as *E. armatus, E. fitchii, E. perbellus, E. baileyi.* In recent years, however, botanists have considered them to be mere varieties of *E. reichenbachii* and they now even consider *E. caespitosus* and *E. purpureus* to be one and the same as *E. reichenbachii.*

E. reichenbachii grows singly or forms clumps. The stem is up to 30 cm (12 in) high and 10 cm (4 in) across, and is divided into nearly twenty regular ribs. The pectinate spines are whitish-pink, pale brown or dark brown. The flowers are pink to deep reddish-violet. Of the pectinate echinocerei this is the easiest one to grow: it need not be grafted. It likes a warm, sunny situation in summer and a lower winter temperature, even below freezing point.

3

The flowers of *Echinocereus reichenbachii* (1) are rather small at first, becoming larger on the second or third day. In some specimens they may measure up to 12 cm (4¾ in) across. Var. *fitchii* is the easiest to grow and produces flowers most readily. Var. *baileyi* (2) is the most beautiful. The flowers, about 6 cm (2⅓ in) across, are produced by seedlings less than 3 cm (1 in) high. From the time when *E. baileyi* was considered to be a separate species, there exist several varieties. These, however, are of importance only from the collector's, not

the botanist's, viewpoint. Specimens
with white spines are listed as var.
albispinus, the ones with yellow spines
as var. *flavispinus*, those with pink spines
(3) as var. *roseispinus*, and the ones with
brown spines as var. *bruneispinus*.

2   1

Echinofossulocactus albatus (A. Dietr.) Britt. et Rose

The generic name *Echinofossulocactus*, meaning prickly, furrowed cactus, was considered to be too long by some cactus growers, who tried to replace it with a shorter, more practical name. However, according to the rules of nomenclature, the name of a genus cannot be changed for such a reason and so the new name, *Stenocactus*, which began to be used and is still widely used by some cactus gowers, is invalid. Most species of *Echinofossulocactus* are very similar and it is often difficult to identify them correctly. Their many synonyms make this even more complicated.

E. *albatus* is considered a mere variety of E. *vaupelianus* by some botanists. However, it is still listed as E. *albatus* in the catalogues of horticultural establishments and is also thus labelled in amateur collections, so we can adhere to the commonly used name for the time being. E. *albatus* is a globose cactus found growing on the plateau in San Luis Potosí state in Central Mexico. It is coloured blue-green and has a relatively large number of wavy ribs, which, as in most echinofossulocacti, become more numerous with age. The plant's main attraction is the four golden-ochre central spines, each up to 5 cm (2 in) long, which make the cactus look like a prickly golden ball. The radial spines, about ten in each areole, are arranged like rays; these are coloured white and are much shorter and more slender. The flowers, which gave the species its name, are yellowish-white to white. The buds, however, are yellow-green. E. *albatus* does not produce offshoots and is therefore propagated from seed. The seedlings grow well from the start and bear flowers around the sixth year. They must, of course, be kept in a light position in winter or moved to their summer location by mid-spring.

2

1

In youth *Echinofossulocactus albatus* (1) bears only one or very few flowers. Mature specimens produce a far greater number. Several species bear a marked resemblance to the cactus illustrated. One such is the yellow-spined *E. ochoterenaus* (2). The two are difficult to tell apart at first, but only until they produce flowers, for those of

E. ochoterenaus, also appearing in early spring on the crown, are up to 5 cm (2 in) across and the perianth segments are coloured reddish-violet. *E. ochoterenaus* is easy to grow, even for beginners. It requires a well-draining substrate and a sufficiently sunny position.

53

Echinofossulocactus coptonogonus (Lem.) Lawr.

It is often said that the members of the genus *Echinofossulocactus* can be reliably differentiated into two groups: *E. coptonogonus* and the rest. *E. coptonogonus* is the only species that can be positively identified at a glance by its straight ribs, which are triangular in section, and which distinguish it from the other species. These ribs are reflected in the specific name, *coptonogonus,* meaning sharp-edged. *E. coptonogonus* grows in various localities in central Mexico, in Hidalgo, San Luis Potosí and Zacatecas states, on rocky knolls only partly shaded by grass. The stem is globose, 7—10 cm (2¾ — 4 in) high, and coloured blue-green. It has only ten to fourteen broad, robust ribs up to 1.5 cm (½ in) high. There are usually five upward-curving spines in each areole, but sometimes only three. One of these, the longest, is up to 3 cm (1 in) long and quite flat. The pale violet flowers are approximately 3 cm (1 in) long and 4 cm (1½ in) across. *E. coptonogonus* is more prone to root rot than the other *Echinofossulocactus* species and therefore should not be watered until the substrate is thoroughly dry. In summer it should be placed in a warm spot that is not too shaded, and in winter it should be provided with a temperature of 8—15°C (46—59°F). If the plant is overwintered in the dark, it should be moved to its summer location early in spring so that it will form buds. If moved later it will not flower.

2

Echinofossulocactus coptonogonus (1) begins to turn brown and corky at the base at a fairly advanced age. This change in the colour of the skin is quite natural in all species of *Echinofossulocactus.* The specimen that was first described had five spines to each areole. Later, however, specimens with only three spines to each areole were found in several localities. In their natural habitat, for example in San Luis Potosí, as well as in cultivation in collections, the three spines are a constant characteristic in their

progeny. Another characteristic by
which these plants differ from the
specimen originally described is their
size; the stem may be up to twice as big
and reaches a height of about 15 cm
(6 in). For this reason they were
classified as *E. coptonogonus* var.
maior (2).

1

Echinofossulocactus multicostatus (Hildm.) Britt. et Rose

Adult specimens of *E. multicostatus* may be readily identified by the large number of slender, wavy ribs, from which the plant also takes its name. Like the other members of this genus, it is found only on the Mexican plateau, in Coahuila, Chihuahua and Durango states, where it grows in intermittent insular localities on rocky calcareous soil. Usually it occurs as a solitary specimen. The globose, flat-topped body is up to 10 cm (4 in) across in adult plants and coloured a vivid green. It usually has 80 to 120 ribs, very occasionally somewhat fewer. There are only two or three areoles on each rib, which are covered with white felt in young plants. Each has four to six radial spines, which point downwards, while at the top of the areole three central spines point upwards. The spines are markedly variable in length. In some specimens they are approximately 4—5 cm ($1\frac{1}{2}$—2 in) long, in others 8—10 (3—4 in) or even 12 cm ($4\frac{3}{4}$ in). They also exhibit great variability in curvature (ranging from straight to greatly curved) as well as colour (ranging from white through brown to nearly black). The flowers, which are not very striking, measure approximately 2.5 cm (1 in) in diameter and, as in other members of this genus (with the exception of *E. phyllacanthus*), are produced only once a year, in early spring. *E. multicostatus* grows well on its own roots and is grafted only when more rapid attainment of the plant's typical appearance is the desired object. The plant requires a sunny, well-ventilated location and a winter temperature of about 10°C (50°F).

2

Echinofossulocactus multicostatus (1) has more ribs than any other cactus. Adult specimens have three to four ribs per 1 cm ($\frac{1}{3}$ in) of their circumference. Though young plants grow rapidly, it takes many years before they acquire the typical appearance of adult specimens. Five-year-old seedlings (2), for example, have only about 30 ribs, ten-year-old individuals about 60, and not until a well-advanced age does their number

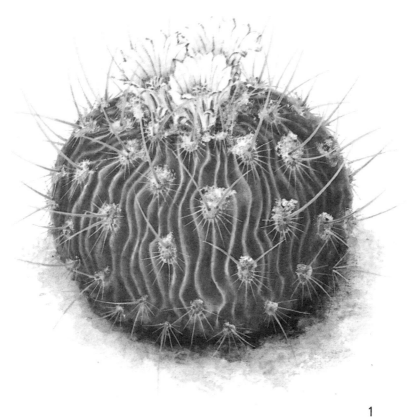

1

3

exceed 100. For this reason, small seedlings are always difficult to identify. Not only the spines but also the flowers of *E. multicostatus* exhibit marked variability. The inner perianth segments may be white with a violet stripe down the middle (3) or completely reddish-violet.

Epiphyllum ackermannii (HAW.)

The plants most widely cultivated under the name *Epiphyllum acker-mannii* are the hardiest red-flowered hybrids of this species with char-acteristics comparable to those of commonly grown houseplants. The *Epiphyllum* species found growing in the wild, more often called by the apt, but invalid, name *Phyllocactus,* i.e. leaf cactus, are much more delicate and therefore soon ceased to be grown in window gardens. Cultivated nowadays are a great many beautiful hybrids, developed with the aim of obtaining large-flowered forms, forms with flowers of unusual colours, multiflowered forms, double forms, long-flowering forms, etc. for which purpose epiphyllums were often hybridized with other related genera, such as *Heliocereus* and *Selenicereus.* They are extremely popular houseplants that should be grown in a nourishing substrate. In summer they may be put outdoors. They do best in a lo-cation sheltered from severe winds and at least moderately shaded. A period of vigorous flowering in spring is followed by a resting per-iod at summer's peak, when watering should be limited and when the plants may also be transplanted. In winter, they should be kept in a cool room at a temperature of 5—12°C (41—54°F) and the substrate should be kept moderately moist. The plants may be propagated throughout the entire growing period by means of cuttings.

E. ackermannii is native to the tropical region of Mexico, Chiapas and Oaxaca states. It is an epiphyte that grows in the humus that has collected on the trunks or in the forks of the branches of broad-leaved trees, where it forms huge clumps. These are composed of dozens of long (*c.* 0.5 m/1½ ft), greatly flattened or three-winged stems resembling leaves. Large crimson flowers grow from the areoles on the wavy edges of the stems.

2

No other cacti were hybridized with such effort and success as the *Epiphyllum* species. Loveliest are the crimson-flowering hybrids (1), developed in the early nineteenth century by crossing *E. ackermannii* and *Heliocereus speciosus*. *Epiphyllum* species are cultivated mainly for their flowers, which may definitely be said to be among the most beautiful in the world.

Nevertheless, some species are also interesting for the unusual shape of their stems. One such is *E. anguliger* (2), with deeply cut, flattened stems.

1

Epithelantha micromeris (Engelm.) F. A. C. Weber

Epithelanthas are typical miniature cacti of North America. They differ from mammillarias, to which they bear the closest resemblance, by the location of their flowers. Whereas in mammillarias the flowers form a wreath near the crown, in epithelanthas they emerge from the newest areoles on the crown. Epithelanthas are cacti with narcotic effects but they contain fewer hallucinogens than *Lophophora williamsii*, the most familiar of the narcotic cacti. *E. micromeris* has a sporadic distribution in northern Mexico and southern Texas, where it is found in very sunny locations, on calcareous knolls as well as in dry riverbeds. It has several varieties, some of which were originally classified as independent species. Var. *micromeris* grows singly as a rule. It attains a diameter of approximately 4 cm (1½ in) and each areole bears 20—30, mostly snow-white, marginal spines. Var. *bokei* has the greatest number of spines per areole (*c.* 40). Var. *greggii* produces numerous offshoots and forms huge clumps. The spines jut from the stem in every direction and are variable in coloration. Characteristic of var. *pachyrhiza* is the turnip-like root and short, delicate spines. Epithelanthas are hard to grow on their own roots and are therefore usually grafted. Plants with their own roots should be watered only when the substrate is thoroughly dry, because frequent watering causes root loss. In summer, the plants require a warm situation and full sun; in winter a temperature of 5—12°C (45—54°F).

2

Epithelantha micromeris (1) and its varieties have small, pale yellow flowers that are about 5 mm in diameter and generally autogamous. They are produced repeatedly throughout the summer. The fruits (2) are only about 12 mm (½ in) long and contain relatively large seeds. The red colour of the pulp makes them more attractive than the inconspicuous flowers. The varieties are distinguished primarily by their spines. However, the differences are clearly visible only when viewed with a magnifying glass. Whereas in var.

1

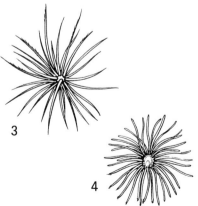

3

4

micromeris (3) practically all the spines
in the areole are the same length, in var.
unguispina the lower spines are
markedly longer, and in var. *bokei* (4)
the spines in the areole are shorter and
very numerous.

Ferocactus glaucescens (DC.) Britt. et Rose

Cacti of the genus *Ferocactus* are typical of the xerophilous vegetation of Mexico and the south-western USA. They are the largest of the globose cacti and often a dominant feature in botanical garden collections as well as in the glasshouse collections of amateur cactus growers.

F. glaucescens may weigh as much as several tens of kilograms in the case of old specimens. It is striking even as a small seedling because of the beautiful blue-grey-green colour of its skin (from which it derives its name) and its amber-yellow spines. In its native habitat of central Mexico it grows mostly at lower elevations (up to 1,500 m / 5,000 ft) on calcareous substrates. It is either solitary or else produces offshoots. The stem is globose to cylindrical, 50—60 cm ($19\frac{1}{2}$—$23\frac{1}{2}$ in) across, 45—70 cm ($17\frac{3}{4}$—$27\frac{1}{2}$ in) high, and bluish. The spines are yellow, 2.5—3.5 cm (1—$1\frac{1}{3}$ in) long, and nearly straight. There are usually four to eight marginal spines and one central spine in each areole, but the latter may be absent. The flowers are yellow, about 4 cm ($1\frac{1}{2}$ in) long and 2.5—3.5 cm (1—$1\frac{1}{3}$ in) across. On plants grown in cultivation, they emerge early in spring and remain open for three to four days. *F. glaucescens* is one of the loveliest and at the same time one of the most easily cultivated *Ferocactus*. In a warm and sunny aspect it tolerates frequent and liberal watering during the growing period. If the substrate is sufficiently nourishing, it produces flowers around the twelfth year, when it measures about 15 cm (6 in) in diameter. In winter, it requires a temperature of more than 10°C (50°F); lengthier exposure to lower temperatures causes fungal diseases of the skin.

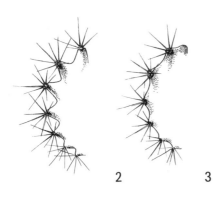

2   3

Ferocactus glaucescens (1) is a very ornamental plant, although its flowers are rather inconspicuous and open only in sunny weather. The closely related *F. schwarzii* grows in the relatively damp forests of the Western Sierra Madre mountain range on the western coast of Mexico, where it grows together with mosses, lichens and appressed orchids in rock crevices and on the inaccessible ledges of river gorges. The small seedlings, like those of *F. glaucescens*, are covered with prominent mammillae (2), which later merge to form ribs (3). In adult specimens there are fewer

1

4

spines in each areole and a continuous felted border (4) forms on the edge of the ribs. Around its twelfth year, the plant begins to bear flowers; these are yellow and measure about 4 cm (1½ in) across. Old plants attain a width of 50 cm (19½ in) and a height of 80 cm (32 in). Cultivation is the same as for *F. glaucescens*.

Ferocactus latispinus (HAW.) BRITT. ET ROSE

W. N. Haworth described this cactus in 1824 as *Cereus latispinus*. It was not until 1922 that it was put in the genus *Ferocactus* by the American botanists L. N. Britton and J. N. Rose, who had newly established this genus. Translated, *Ferocactus* means 'formidably or savagely spined cactus', and this truly reflects the character of all members of the genus. *F. latispinus* is found in central Mexico at elevations of approximately 2,000 m (6,500 ft), where it grows on grasslands and rocky knolls. The body is flat-topped, globose or flat, up to 40 cm (15¾ in) across and 30 cm (12 in) high. The nine to fifteen radial spines are extremely variable, ranging from dark, thick spines to ones that are pale and needle-like. In each areole there are four central spines, one of which points downwards and is curved or hooked. The flowers, up to 4 cm (1½ in) across, are generally purplish-red, but sometimes yellow. Although *F. latispinus* grows on calcareous substrates, it is not necessary to add limestone rubble to the growing medium. What is necessary, however, is that the substrate be free-draining and nourishing. Do not be in a hurry to start watering in spring. Like all ferocacti, *F. latispinus* begins growth somewhat later and will itself reliably indicate when this occurs by the emergence of glowingly coloured spines that are soft and break off easily at first. This rarely takes place before the end of May. During the growing period, apply water at intervals only in congenial weather. This cactus requires a sunny and warm situation and in winter conditions that are not too cool and a lower humidity (atmospheric moisture), otherwise unattractive spots, caused by a fungal disease, appear on the skin.

2

Ferocactus latispinus (1) has beautiful and interesting central spines. The three uppermost spines are straight, up to 4 cm (1½ in) long and point upwards. The fourth, lowermost spine is curved or hooked, prominently transversely striped, up to 5 cm (2 in) long, often very wide, and, unlike the other three, points downwards. The unusual width of the spines, which may be up to 1 cm (⅓ in), inspired the plant's name (*latispinus* means wide-spined). The spines are ruby-red, as well as yellow to

1

3

white (2). The flowers may also be dark carmine to yellowish. Native to Puebla and Oaxaca states is the variety *spiralis*, also known as *Ferocactus recurvus*. It has a more elongated stem and its ribs are not as spiral in shape. There are fewer radial spines in the areole (five to seven) and the central hooked spine is up to 6 cm ($2\frac{1}{3}$ in) long (3).

Glandulicactus uncinatus (Galeotti) Backeb.

This species is generally found in collections under the name *G. uncinatus*, even though many taxonomists class it in the genera *Ancistrocactus, Hamatocactus* or *Ferocactus*. Like all members of this genus, *G. uncinatus* has glands in the areoles that secrete a sweet viscid liquid that is attractive to ants, which build their anthills nearby and live in a partially symbiotic association with the plant. *G. uncinatus* provides the ants with sweet juice and in return the ants rid it of parasites. *G. uncinatus* has an interrupted distribution in northern Mexico and the southernmost states of the USA. The stem is slightly elongated in age, up to 20 cm (8 in) high, greyish-blue-green, and does not produce offshoots. The ribs are composed of prominent tubercles that enable the plant to shrivel during the resting period by as much as 30 per cent of its original volume. At this time it is thickly covered by a mass of huge spines, the longest of which are conspicuously hooked at the tip. It is these that gave the plant its name — *uncinatus* means hooked. The flowers are also variable, ranging from pale cinnamon-red through brownish-violet to black. *G. uncinatus* is not easy to cultivate and hence cactus growers often favour grafting. Though their growth is reliable, grafted specimens grow too tall and produce fewer spines. The best method, therefore, is to cultivate it on its own roots. It requires a substrate poor in organic substances, careful watering and a warm, sunny situation.

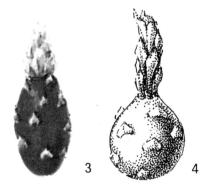

3   4

In its native habitat *G. uncinatus* (1) forms numerous isolated populations separated not only in terms of distance but also by mountain ranges. Found in the separate localities are different local forms that apparently became further differentiated in the course of intensive development. In some localities they are small in size, in others they are relatively large. Some have short spines, others unruly spines arranged every which way, with prominent, hooked central spines up to 10 cm (4 in) long (2). Southern populations (from San Luis Potosí,

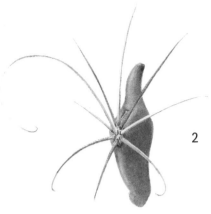

Nuevo León, Zacatecas and Coahuila) are classified as *G. uncinatus* var. *uncinatus*; northern populations (from Texas, New Mexico and Chihuahua) as *G. uncinatus* var. *wrightii*. They can be reliably distinguished by their fruits. Var. *uncinatus* has fleshy, club-shaped fruits (3) with about 300 relatively large seeds; the fruits of var. *wrightii* are more or less globose (4) and contain fewer seeds.

2

1

Hamatocactus setispinus (ENGELM.) BRITT. ET ROSE

H. setispinus is a good cactus for the novice cactus grower. The seedlings grow rapidly and even two- to three-year-old seedlings will generally bear relatively large, pale yellow flowers. In ensuing years, *H. setispinus* produces buds continually throughout the growing period, in other words from early spring until the autumn. It is distributed over a relatively large range; separate localities are found in southern Texas and in Coahuila, Nuevo León and Tamaulipas states in Mexico. In young specimens the stem is globose, in old plants, however, it becomes slightly lengthened and at a diameter of approximately 10 cm (4 in), it reaches a height of 15 cm (6 in). Besides slender, radiating marginal spines, each areole has one central spine that is 2—4 cm ($\frac{3}{4}$—$1\frac{1}{2}$ in) long and hooked at the tip. The flowers emerge near the crown. They are about 7 cm ($2\frac{3}{4}$ in) long and have a pleasant, delicate fragrance. The perianth segments are pale yellow, changing to carmine at the base and thereby forming a regular ring in the centre of the flower. The colour of this ring is quite variable; in some specimens it is pale, almost pink, in others deep red. Because *H. setispinus* grows singly, it is propagated only by means of seeds. It requires a mineral substrate, an adequately warm and sunny situation and watering that is regularly spaced throughout the whole growing period. Like most cacti, it may be overwintered in the dark without any diminishing of flowering in the following year. In winter, it should be kept in an absolutely dry substrate at a temperature of 5—15°C (41—59°F).

2

Hamatocactus setispinus (1) is a cactus that shows the greatest willingness to produce flowers with the greatest frequency. As one series of flowers fades, new buds are already being formed. The flowers (2) develop quite rapidly, especially prior to opening, when they increase up to 5 cm (2 in) in length within a single week. *H. setispinus* is an autogamous species. It produces globose, bright red fruits about 1 cm ($\frac{1}{3}$ in) across even when the pollen from the anthers is transferred to the stigma

1

3

of one and the same flower. The fruit
contains approximately 150 quite large
seeds. *H. setispinus* is a very variable
plant exhibiting marked differences in
the length and coloration of the central
spines. Several varieties have been
described on the basis of these
differences; most striking is var. *orcuttii*
(3) with pale yellow central spines.

Hylocereus undatus (HAW.) BRITT. ET ROSE

The genus *Hylocereus* comprises climbing cacti noted for their large, white, nocturnal flowers with a heady, penetrating scent. These cacti are distributed throughout Central America and even further south in South America. The most widely cultivated of all is *Hylocereus undatus,* which is found growing in eastern Mexico and the Greater Antilles. It belongs to the group of so-called climbing cerei. The seeds germinate in the ground, sometimes at the foot of solitary trees or in rather thin deciduous woods. The main stem rapidly climbs up the trunk until it reaches a spot that receives sufficient sun. There it forms a huge branched clump of three-winged, practically bare, non-prickly stems up to several metres long. Produced at the ends of the stems are large white flowers, about 25 cm (10 in) across, that last only one night. *H. undatus* grows very rapidly and is therefore a suitable plant for larger glasshouses and heated verandas and as a covering for rear walls that are at least partially protected from full sun. In horticultural establishments, it is often used as rootstock for grafting — chiefly for coloured cultivars without chlorophyll. Plants grafted on such stock show excellent growth. In winter, they require a light situation, a temperature above 15°C (59°F), and occasional watering. *H. undatus* requires a humus-rich substrate, preferably with an admixture of peat, and liberal watering together with a liquid feed throughout the growing period. This applies also to stock on which other cacti are grafted. Propagation is by means of stem cuttings.

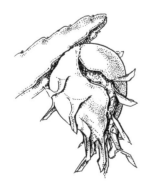

2

Hylocereus undatus (1) is very popular in the tropics for its large, glowing white flowers. It is planted in parks and gardens as a climber for covering walls, arbors, frameworks and the like. *H. undatus* is prized not only for aesthetic reasons but also for its juicy fruits, about the size of an apple (2), produced after the flowers have been pollinated by hawk moths. These fruits are very tasty and are a favourite fruit in Mexico and the Antilles. As in all other climbing cacti, the long, bare, spineless stems of this species are furnished with clinging roots by means of which the stem attaches itself to its support (3).

1

3

Leuchtenbergia principis Hook.

L. principis is the only member of the monotypic genus *Leuchtenbergia*. It occurs rarely on calcareous substrates on the central Mexican plateau and is currently one of the most greatly endangered cacti in the wild. It reaches a height of 60—70 cm ($23\frac{1}{2}$—$27\frac{1}{2}$ in) (slightly less in cultivation) and has straight, triangular mammillae up to 15 cm (6 in) long and covered with a bluish bloom. The one or two central spines in each areole are greatly twisted, papery, and 12—15 cm ($4\frac{3}{4}$—6 in) long. They break off easily. In late summer, buds covered by scales appear on the top of the youngest mammillae. Within the period of one month these will develop into yellow, delicately scented flowers, 8 cm (3 in) in diameter. They bloom for about a week; in cooler weather even longer. *L. principis* is propagated only by means of seeds. The large seeds germinate well if sown in a rather coarse, thoroughly moist substrate and pressed in. When they have germinated, drain off excess water and, about a month later, begin airing the interior of the box, for the seedlings are relatively susceptible to damping-off, often brought on by soil that is permanently excessively wet. Older plants should be put in a mineral, somewhat loamy, substrate and watered at shorter intervals. Whereas permanently wet soil over a lengthier period causes root rot, the drying out of the soil during the growing period results in the drying of the tops of the mammillae and an unattractive appearance.

3

Leuchtenbergia principis (1) adapted to excessive sunlight and lack of moisture in a slightly different manner than most common cacti. Division of the stem into numerous long mammillae resulted in a sort of leaf succulence. Because such mammillae increase the surface area, not only for photosynthesis but also for evaporation, the plant was forced to develop mechanisms that would prevent its complete dehydration during the resting period. At this time, the

1

mammillae contract in plants filled with water (2), thereby minimizing the area exposed to the sun. Flowers are formed at the tips of the mammillae, where they are replaced by fruits (3) following pollination. These contain about 250 seeds that are unusually large for cacti.

2

Lophophora williamsii (LEM. EX SALM-DYCK) COULT.

Some cacti contain alkaloids that cause hallucinations when ingested. Examples of cacti with hallucinogenic or narcotic effects are species of *Ariocarpus, Epithelantha, Obregonia* and *Pelecyphora*. The most familiar of such cacti, however, belongs to the genus *Lophophora*. The narcotic effects of *L. williamsii* were known to the Aztecs, who made use of them in their rituals and religious ceremonies. Nowadays the plant's alkaloids, chiefly mescaline, are used in the treatment of certain mental diseases. *L. williamsii* grows in the dry regions of southwestern Texas and northern Mexico. During the resting period it is almost entirely drawn down into the ground, the stems rising above the surface only during the growing period. Their collection is strictly prohibited. The spineless bodies grow singly or in clumps. The underground extension is a massive turnip-like tap root that is often bigger than the above-ground parts. *L. williamsii* is quite variable in the arrangement of its tubercles and ribs, which has resulted in the introduction of many unjustified varieties. Specimens with numerous ribs are designated as var. *pluricostata*, specimens with five ribs (their number generally increases with age) as var. *pentagona*, specimens with ribs that break up in age into flat hexagons as var. *decipiens*. *L. williamsii* should be grown in a deep container in a sandy-loamy substrate. Propagation is by means of seeds. The fruits appear nine to twelve months after self-pollination.

2

Lophophora williamsii (1) has a small, blue-green, flat-topped, globose body with about eight faintly bumpy ribs. Tufts of hairs grow from the areoles on the ribs. The flowers emerge from the areoles near the crown and are produced throughout the summer. They are 1—2.5 cm ($\frac{1}{3}$—1 in) long and generally coloured pale pink. In some specimens the flowers are a deep pink (2) or violet-red. These are characteristic, for example, of *L. williamsii* var. *jourdaniana*, often classed as an independent species. It also differs by having several short spines in each

areole, that remain on younger plants and offshoots for many years. Another readily distinguished species of this genus is *L. diffusa* (3), in which adult specimens have diffuse ribs. It is found in isolate localities in Querétaro state in Mexico. At first it grows as a solitary specimen; later it forms clumps. The flowers are white, pinkish or yellowish and are not autogamous. It also differs from *L. williamsii* in the composition of its alkaloids.

3

1

Mammillaria bocasana POSELG.

The genus *Mammillaria* is one of the largest in the Cactaceae family. Although several hundred species were described in the past, nowadays the consensus of botanical opinion is that there are about 200 species. These are distributed throughout central and northern Mexico and the south-western USA. They differ from other cacti by having their mammillae distributed all over the surface instead of being arranged in the form of ribs, and in the way that their flowers are produced. Whereas the flowers of most cacti grow from the areoles, in mammillarias they grow from the axils, i.e. from the hollows between the mammillae.

M. bocasana is one of the commonest mammillarias in collections because of its attractive appearance, rapid growth and ease of propagation. It is native to Sierra de Bocas in the state of San Luis Potosí. It grows singly at first, but later forms large clumps of spherical stems 4—5 cm (1½—2 in) across. As in most mammillarias, the spines are quite variable. One of the central spines is hooked, 5—8 mm long, and coloured yellow-brown or yellowish. Specimens with pure yellow spines were described as var. *flavispina*. From the botanical viewpoint, however, this is not a well-founded designation. It is the delicate, thick, hair-like radial spines, almost entirely concealing the mammillae, that make *M. bocasana* so attractive. There are 25—30 of them to each areole. They measure up to 2 cm (¾ in) in length and are outspread at first, later greatly intertwined. Plants with the longest and greatest number of hair-like spines were described as var. *splendens*. This variety is likewise of importance purely from the collector's aspect, not botanically. *M. bocasana* is easy to cultivate, grows rapidly and is readily propagated by offshoots. To obtain a thick covering of hair and compact clumps of stems, apply water lightly and only occasionally during the growing period, and in winter keep the substrate absolutely dry.

Mammillaria bocasana (1) has a globose body and forms large clumps composed of dozens of such spherical stems, 4—5 cm (1½—2 in) in diameter. The flowers are small, 1.5 cm (½ in) across, the perianth segments creamy-white with a central reddish stripe. The flowers readily escape notice amid the thick hair-like spines. A more striking ornament is the fleshy, bright red fruits (2). Also grown in collections is

4

1

3

2

M. bocasana var. *roseiflora* (3), which has
pink flowers. This, however, is not
a true variety but a hybrid, obtained by
crossing *M. bocasana* with the
violet-flowered species *M. zeilmanniana.*
Also widely cultivated is the mutant
M. bocasana var. *multilanata* (4), so
named for its dense cover of spines,
which are shorter and often number
as many as 200 to each areole. This is
one of the most beautiful mammillarias
of all.

Mammillaria candida SCHEIDW.

M. candida grows in several states in Mexico (Tamaulipas, Nuevo León, Coahuila and Zacatecas) on limestone rock in open country with sparse scrub, together with other xerophilous plants such as *Agave, Hechtia, Yucca* and the like. It is very sparsely distributed throughout this wide range and is therefore pollinated in an unusual manner. To enable the pollination of plants separated by great distances, it flowers earlier than most other cacti, bearing flowers at a time when the woody plant *Prosopis juliflora,* common in these parts and much visited by various hymenopterous insects, is still in full bloom. These insects will then pollinate even lone, isolated cacti. *M. candida* (which translates as dazzling white) has a flat-topped, globose stem, later becoming slightly lengthened, that measures about 10 cm (4 in) in diameter. In each areole there are eight to twelve central spines, 5—9 mm long and coloured white, although often pinkish to brownish at the tip, and about 50 radial spines, likewise coloured white and arranged in several rings. The flowers are small, approximately 2.5 cm (1 in) across, and pinkish. The smooth, glossy black seeds are produced only very occasionally in comparison with other mammillarias and for that reason this species is generally propagated by means of offshoots. In cultivation it requires a fair amount of sunlight, a mineral substrate, and the careful application of water only after the substrate is completely dry. The only closely related species is *M. ortizrubiona.* It is very difficult to distinguish between the two. The latter, however, has larger and more loosely arranged mammillae and a smaller number of spines.

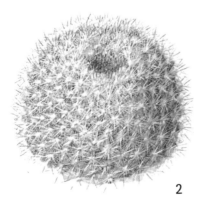

Mammillaria candida (1) forms sideshoots only as an adult plant when the stem reaches its maximum diameter and begins to lengthen slightly. The dense covering of pale spines protects the skin from direct sunlight and enables the plant to grow even in unshaded locations in its native land. The central spines are either entirely white or faintly yellowish; in some specimens they are red or brown towards the tip, giving the plant a darker pinkish hue. Specimens

2

with such darker coloured tips are named var. *rosea* (2). These, however, must be viewed only as a deviation within the natural variability of the species. They occur sporadically throughout the species' entire range and are therefore of importance only to the collector.

1

M. carmenae was described in 1953 but it did not appear in collections and botanists searched for it intensively in ensuing years, without success. Not until 1976 was this species discovered anew in Mexico by Dr A. B. Lau, on the practically inaccessible northern slopes of the Eastern Sierra Madre, not far from the town of Jaumave. *M. carmenae* is a typical stenoendemic species, i.e. one restricted to a very limited locality. It is found in a small area, on granite rock faces surrounded on all sides by practically impenetrable pine stands. There it grows on north-facing rock terraces in rather thick layers of humus, more often in grass and amid pine needles beneath scattered pine trees, as well as in rock crevices together with mosses, ferns and lichens.

M. carmenae is a miniature plant covered with a thick coat of radial spines. It occurs singly or in groups. The downy, marginal spines are very numerous, as many as 100 in a single areole, but there is no central spine. The flowers are small, about 1 cm ($\frac{1}{3}$ in) across, and coloured pinkish-white. Even after it was rediscovered *M. carmenae* continued to be a great rarity and was thus generally grown as a graft. On the whole, it is sufficiently common nowadays but many cactus growers still prefer to graft it, primarily on *Eriocereus jusbertii* stock, on which it grows rapidly and produces flowers in its second year. In a rather acidic mineral substrate it grows well even on its own roots and attains sexual maturity relatively quickly, bearing flowers in the third and fourth year.

2

Found growing together in the same locality are specimens with golden-yellow spines (1) as well as ones with yellowish-white spines (2). The surface area of each spine is increased by its pubescence. Similar, but even more pubescent, are the spines of *M. plumosa* (3), which grows about 100 km (62 mi) further north. Because of its delicate, white, non-prickly, feathery spines (4), it is often commonly called the 'powder-puff'; however, its Latin name, *plumosa*, means feathery. In

1

3

4

collections it seldom bears flowers, and
when it does then generally only in
winter. For that reason it is usually
propagated by means of offshoots,
which are readily produced around its
periphery. Like *M. carmenae*, it requires
a mineral, well-draining substrate,
careful watering and a sunny aspect.

Mammillaria geminispina Haw.

M. geminispina is one of the loveliest of the so-called white mammillarias of the Mexican plateau. In Hidalgo, San Luis Potosí and Veracruz states, it grows on steep limestone slopes, in clefts, gorges, canyons and deep valleys carved in the underlying limestone rock by river systems over the ages. At first it grows as a solitary specimen, but later it produces numerous offshoots and forms huge clumps composed of several dozen heads. The individual stems may grow up to 20 cm (8 in) high and 10 cm (4 in) across and when bruised they exude a milky fluid. The spines of this species are very variable. There are approximately twenty shorter, chalk-white, radial spines and two, (very occasionally as many as six) central spines. It is the latter that gave the species its name — *geminispina* means bearing spines in twos. The spines vary not only in terms of length but also in the degree of curvature of the central spines, which may be nearly straight or greatly twisted and intertwined. In some specimens they are entirely white, in others they have variously dark tips. The flowers are approximately 2 cm ($\frac{3}{4}$ in) long and coloured reddish-violet. *M. geminispina* has a relatively delicate root system that is quite susceptible to rot. It should be grown in a mineral substrate and watered when the soil is thoroughly dry.

2

Besides the reddish-violet flowers, the most attractive feature of *Mammillaria geminispina* (1) is its white central spines, which may be quite long in some specimens. Plants with extremely long spines were named *M. geminispina* var. *nobilis* (2), which means noble, stately or comely. Botanically, the introduction of this variety is not justified. The similar and closely related species

1

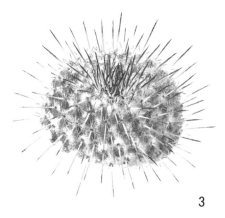

M. parkinsonii (3) likewise forms huge
clumps produced mainly by
dichotomous division, i.e. the division of
individual heads. The colour of the
central spines is extremely variable, as is
their length. They may be chalk-white,
brown or nearly black. Whereas
M. geminispina has reddish-violet
flowers, those of *M. parkinsonii* are
straw-coloured and quite inconspicuous.

3

Mammillaria hahniana WERDERM.

M. hahniana is native to the states of Guanajuato and Querétaro in Mexico, where it grows on the plateau on calcareous substrates at elevations of nearly 2,000 m (6,500 ft). As a rule it forms rather large clumps; only very occasionally does it remain solitary. The body is globose, later oblong, up to 10 cm (4 in) across, and exudes a milky fluid when bruised. It is covered with numerous mammillae. The central spine is needle-like and whitish with a reddish-brown tip. There is generally only one in each areole, only rarely are there two to four. The long, white, hair-like radial spines are 5—15 mm long and, according to the original description, number 20—30, although sometimes there are more. A typical characteristic of this species is the axillary hair-like bristles, which are very long (over 4 cm/$1\frac{1}{2}$ in) and numerous in some specimens, whereas in others they are short and sparse. Most highly prized by cactus growers are the selectively bred forms with long white 'hairs' that may completely cover the plant. Numerous purplish-red flowers emerge early in spring near the crown to form a wreath. Two varieties — var. *giseliana* and var. *werdermanniana*—that differ from the type species in the number and length of the axillary bristles and radial spines, have been described, although there is no justification for this from the botanical viewpoint. On the other hand, some mammillarias, originally described as independent species, that might justifiably be classed as varieties or forms of *M. hahniana* include *M. mendeliana, M. woodsii* and *M. bravoae. M. hahniana* is comparatively easy to grow and produces flowers readily. It has no special requirements but produces its loveliest spines in a sunny situation.

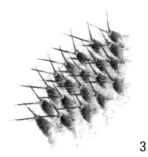

3

The spines, and especially the axillary bristles, of *Mammillaria hahniana* (1) are very variable in number as well as length. Whereas some specimens are completely covered by a dense mass of spines and bristles, others are almost bare of these (2). As already stated,

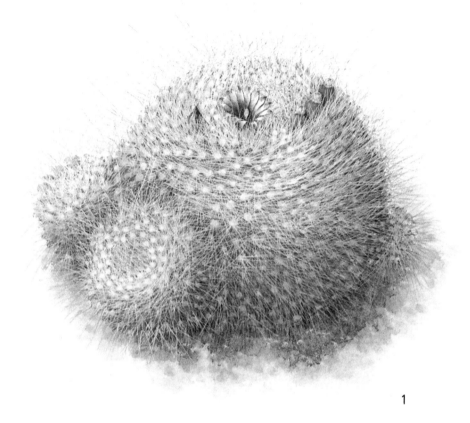

1

2

several mammillarias growing in the same region and classified as independent species may be considered varieties of *M. hahniana.* One such is *M. woodsii* (3), distinguished by dark central spines, a mostly solitary habit and body more than twice as big as *M. hahniana.*

Mammillaria herrerae WERDERM.

M. herrerae is one of the loveliest of the miniature cacti. Its small size makes it ideal for window gardens or other types of collections with limited space. It was discovered more than 50 years ago and continues to be relatively rare due to the difficulties connected with its vegetative propagation (it does not flower readily in cultivation) and also due to the inaccessibility of its native sites. It has only a sparsely scattered distribution, occurring in limestone pockets, filled with gravel and loamy soil, together with grass and xerophilous ferns on sunny slopes in the neighbourhood of Vista Hermosa in the state of Querétaro, Mexico. It either remains solitary or else produces offshoots and forms small clumps. The numerous spines (about 100 in each areole) are 1—5 mm long, gleaming white with a yellowish base, and spread outward in a regular radiate arrangement. The small stems, covered in white spines, merge with the background of the white limestone. This is known as mimicry, i.e. the superficial resemblance of an organism in coloration or shape to objects in its natural environment as a means of protection. During the flowering period, however, it attracts notice by its pale pink to violet flowers, measuring 2—2.5 cm ($\frac{3}{4}$—1 in) in diameter. Though *M. herrerae* is rather difficult to cultivate, it can be grown by novice cactus growers provided it is grown as a grafted specimen. Good stock for this purpose is *Eriocereus jusbertii*, on which it almost retains its natural shape, only producing offshoots somewhat more liberally. To ensure regular flowering, it should be kept in a sunny location and newly forming offshoots should be removed regularly. In winter it should be kept at a temperature that is not too cold, and provided with lower relative humidity, otherwise its skin is often attacked by fungal diseases.

2

Mammillaria herrerae (1) has violet flowers that are relatively large in relation to its miniature size. There is also a white-flowered variety — var. *albiflora* (2) — that is sometimes even classed as a separate species. It differs by having white flowers, fewer spines and a more elongated stem. Another quite similar species, likewise with numerous snow-white spines, is *M. humboldtii* (3). It is a beautiful cactus forming large clumps of heads

completely covered with short white
spines. Compared with *M. herrerae*, it
not only attains larger dimensions but is
also more easily grown on its own roots
and flowers more readily; the flowers
are reddish-violet and measure 1.5 cm
($\frac{1}{2}$ in) in diameter. Although it, too, is
native to Mexico, more precise
information as to its place of occurrence
is kept secret in order to protect this
unique gem from extinction.

3

1

Mammillaria longimamma DC.

For mammillarias with long mammillae the American botanists L.N. Britton and J.N. Rose introduced a new genus in 1923, naming it *Dolichothele* after the long nipples (the Greek word *dolichos* means long and *thele* means nipple). That is why *M. longimamma* is often found in collections under the name *Dolichothele longimamma.* From the current viewpoint, the genus *Dolichothele* is unjustified. *M. longimamma* is native to Hidalgo state in Mexico, where it grows in grass in the shade of shrubs and, during the dry period, draws down quite considerably into the ground. Also native to Hidalgo is the closely related *M. uberiformis,* which differs by having shorter mammillae and darker spines. Because these are the only differences between the two, it would be more suitable to view *M. uberiformis* as a variety or form of *M. longimamma. M. longimamma* has a thick, turnip-like, greatly branched root. The stem, 8—15 cm (3—6 in) high, produces offshoots at the base and may form huge clumps in old age. The cylindrical mammillae, 3—5 cm (1—2 in) long and 1—1.5 cm ($\frac{1}{3}$—$\frac{1}{2}$ in) thick, are very soft, jut outward freely and contain a watery fluid. The flowers grow from the upper axils. They are large compared with the flowers of other mammillarias, measuring 5—6 cm (2—2$\frac{1}{3}$ in) across, and are coloured lemon-yellow. They open fully only in sunny weather. Because *M. longimamma* has a large turnip-like root serving the plant as a reservoir of water during the resting period, it should be put in a deep container. It requires a partly shaded situation because its skin is not protected by prominent spines and the plant could readily be scorched by direct sunlight, particularly in spring when it is not yet in full growth.

2

The mammillae of *Mammillaria longimamma* (1) are extremely variable in length as well as width. Whereas in some specimens they are 2—2.5 cm ($\frac{3}{4}$—1 in) long, in others they may be as much as 7 cm (2 $\frac{3}{4}$ in) long (2). Even though *M. longimamma* has the longest nipples of all mammillarias, many species come close to equalling it in this respect. Easiest to grow is

1

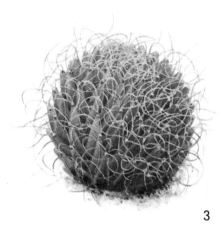

M. camptotricha (3). Its white flowers are quite small, only about 1.5 cm ($\frac{1}{2}$ in) across, but they have a pleasant fragrance. The small bodies are composed of 12- to 14-mm-long nipples terminated by yellow, radial spines that are usually greatly twisted. The variety *albescens*, often classed as an independent species, differs in having white spines that are shorter and straighter.

3

Mammillaria pectinifera (Stein.) F.A.C. Weber

This mammillaria was originally described as *Pelecyphora pectinata*, but is more frequently encountered under the name *Solisia pectinata*. In 1923 L. N. Britton and J. N. Rose put it in a separate genus, which they named in honour of the Mexican botanist O. Solis. However, on the basis of the reproductive organs, for instance the circular arrangement of the flowers near the crown and the formation of the fruits, as well as the amount and composition of milky body fluids, it must be unequivocally incorporated in the genus *Mammillaria*.

M. pectinifera mostly remains solitary; only very occasionally do adult specimens produce a few offshoots. It is found in Mexico in the neighbourhood of the town of Tehuacán in Puebla state. The shape of the body is more or less globose. For a large part of the year it is drawn down below the surface of the ground and only during the rainy season does it protrude 1—2 cm ($\frac{1}{3}$—$\frac{3}{4}$ in) above the surface. This subterrestrial manner of growth enables the plant's small body to survive periods of drought and at the same time limits the possibility of the cactus being discovered during its short growing period, which is conducive to the preservation of the species. Because it is relatively difficult to grow, *M. pectinifera* is generally encountered in collections as a grafted specimen. More experienced growers can grow the plants on their own roots. The seedlings grow rapidly and, as a rule, bear flowers in the fourth or fifth year. They retain their miniature habit and are not deformed by premature growth in early spring. Compared with other common mammillarias, they are more susceptible to damping-off and root rot following ill-advised watering in cooler weather.

3

When grafted on stock, *Mammillaria pectinifera* (1) attains much larger dimensions, flowers more profusely and produces more offshoots. The flowers are 2—2.5 cm ($\frac{3}{4}$—1 in) long and coloured from almost white to deep pink. In their native habitat plants are only 2—4 cm ($\frac{3}{4}$—$1\frac{1}{2}$ in) high and bear fewer flowers. *M. pectinifera* differs markedly from other mammillarias in

1

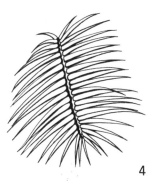

2

5

4

habit. It partly resembles *M. solisioides*, having spines arranged in a ring in the areole (2). It may readily be mistaken for *Turbinicarpus pseudopectinatus* (3) by the novice; the latter, however, bears flowers on the crown and has more and shorter spines. *M. pectinifera* has about 40 spines in each areole (4); *T. pseudopectinatus* about 60 spines (5).

Mammillaria senilis LODDIGES

This lovely cactus is often found in collections under the name *Mamillopsis senilis*. Because of its unusual flowers, it was classed by some botanists in the genus *Mamillopsis*, initially established only for this cactus. *Mammillaria senilis* is native to north-western Mexico where it is found in the Western Sierra Madre region at elevations of 2,400—3,000 m (1,800—9,800 ft). It grows in thin, mixed woods on a partially weathered granitic substrate. In winter, the temperature in these mountains drops below freezing point and *M. senilis* is often covered by snow. Also native to north-western Mexico is the closely related *M. diguetii*, which differs only by having a greater number of rigid, straw-coloured spines and smaller, dark red flowers about 3 cm (1 in) long and 2 cm ($\frac{3}{4}$ in) across. *M. senilis* has a globose body at first, which is later slightly elongated, about 6 cm ($2\frac{1}{3}$ in) across and up to 15 cm (6 in) high. It produces offshoots prolifically and forms many-headed clusters. It is covered with a thick coat of gleaming white spines, always with several hooked central spines in each areole. The large flowers, up to 7 cm ($2\frac{3}{4}$ in) long, have a robust tube and wide-spreading perianth segments with prominent stamens arranged in a bundle around the style. *M. senilis* has a rather delicate root system and for that reason it is recommended that it be grown on stock. If you wish to grow it on its own roots, use a mineral substrate and apply water only occasionally. The seedlings grow rapidly and may flower in the fifth year. However, it sets on flowers reluctantly. *M. senilis* tolerates even winter frosts if kept in a thoroughly dry substrate.

2

Mammillaria senilis (1) differs from other mammillarias in the structure of the flowers. Their colour is also extremely variable, occurring in various shades of orange, red and violet. In collections, growers have succeeded in developing white-flowering specimens (2). This is apparently an abnormality caused by a chance mutation, i.e. loss of the ability to form the characteristic red pigments (betacyanins). The hooked spines aid in the vegetative propagation of the species in the wild. The hooks at the tips of the central spines (3) on the offshoots catch in the fur of animals, whereupon the offshoot is readily pulled off and carried to another place.

3

1

Mammillaria spinosissima LEM.

M. spinosissima flowers readily but is very ornamental even without flowers. It is found in a number of places on the Mexican plateau, for example in the neighbourhood of the capital, Mexico City. The great variability in the colour of its spines is responsible for the fact that several forms have been described as separate independent species (e.g. *M. bella*). *M. spinosissima*, which remains solitary, has an oblong body up to 3 cm (1 in) high and 8 cm (3 in) across. In each areole there are 20—25 radial bristle-like spines, 4—6 mm long and whitish in colour. The number of central spines, according to the original, over 150-year-old description, is supposed to be twelve to fifteen and they should be straight and awl-shaped. Imported plants found in collections generally have fewer central spines — seven to eight, very occasionally as many as twelve. The plant is completely covered with spines and so the name *spinosissima* (which means the spiniest) is an apt one. The flowers are produced in spring. As in other mammillarias, they open in the morning and remain open for about four days. They are reddish-violet, 1.5 cm ($\frac{1}{2}$ in) across, and generally a greater number open at one time, forming a wreath. As the flowers fade, new buds appear so that the plant remains in bloom for several weeks. *M. spinosissima* is the least demanding of cacti. It grows well in a standard substrate in full sun as well as in partial shade. The seedlings grow rapidly as soon as they germinate and bear flowers in the fourth year. The plant's drawback is the overlong shape of the body in old age. To prevent this, at least in part, do not water too often — once a month is sufficient. Although its growth will then be slower, it will remain shorter and will have thicker and more colourful spines.

3

Mammillaria spinosissima (1) is a very variable species in terms of the number, and above all the colour, of the spines. Even though the plant on which the original description is based had dark pink central spines, in collections one will encounter numerous other colour deviations, some of which are very pronounced. Indisputably the most attractive is var. *sanguinea* (2), with

94

1

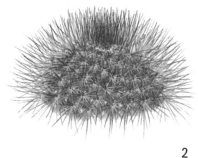

2

spines that are a glowing red when illuminated from behind by the sun. At the other end of the scale are the specimens with light-coloured, sometimes entirely white spines (3). Growers may thus have an interesting and colourful collection composed only of specimens of this one single species.

Obregonia denegrii FRIČ

O. denegrii is the only species in the genus *Obregonia*, established in 1927 by the Czech traveller and cactus collector A. V. Frič. It is found east of the town of Jaumave in the broad Rio Guayalejo valley in the state of Tamaulipas, Mexico, in only two localities that are relatively close to one another. There it grows on shrub-shaded slopes whose calciferous gravelly covering is composed of pure mineral-type soils. It is quite abundant in the respective localities, with approximately two to five plants per square metre. Because *O. denegrii* is an endemic species, and therefore not found anywhere else, it was put on Mexico's list of endangered species in 1981. It has a globose, slightly flat-topped body that in adult plants attains a diameter of 10—15 cm (4—6 in), very occasionally 20 cm (8 in). The stem terminates at the base in a strong, turnip-like root and much of the plant is drawn down into the ground. The areoles at the tips of the triangular mammillae bear two to four weak, non-prickly spines that soon fall. The flowers emerge from the areoles on the crown when the plant measures about 3.5 cm ($1\frac{1}{3}$ in) across and begins to develop a thick white wool on the crown. The flowers are white to pale pink and about 2.5 cm (1 in) across. *O. denegrii* is quite difficult to grow on its own roots. It requires a well-draining, sandy-loamy substrate and watering only in the hottest periods. It is much more easily grown when grafted on *Eriocereus jusbertii* or *Echinopsis eyriesii* stock. The former, however, should be shortened after a time so that the grafted specimen does not look unnatural. In collections, it requires a fair amount of shade. It is susceptible to sun scorch especially in spring, when it is not yet in full growth. At this time, shade it with a sheet of paper placed directly on the plant stem.

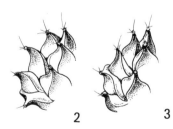

2   3

The stem of *Obregonia denegrii* (1) is covered with scale-like, triangular mammillae flattened on the upperside, up to 2.5 cm (1 in) wide and 1.5 cm ($\frac{1}{2}$ in) long. The mammillae spread out from the body when the plant is filled with water (2) and fold up during periods of drought (3). Found in Tamaulipas state is the similar species *Pelecyphora* (previously *Encephalocarpus*)

strobiliformis (4), which grows at higher
elevations (1,600—1,800 m /
5,250—5,900 ft). The triangular
mammillae, formed only on older plants,
make this species look like a pine cone.
The conditions for growing are the same
as for *O. denegrii*, except that it is
slightly more tolerant of overexposure
to sun.

4

1

Opuntia microdasys (LEHM.) PFEIFF.

Opuntias are the most widely distributed cacti. They are found on the continent of South and North America and the neighbouring islands. The range of their natural distribution extends north into Canada and south as far as the inhospitable region of Patagonia. Some species from cooler regions (for example *O. humifusa, O. rhodantha, O. fragilis*) are grown outdoors throughout the year even in central Europe, for instance in a free-draining rock garden. This, however, does not hold true for one of the most commonly grown opuntias, the species *O. microdasys* from the warm regions of northern Mexico. It forms shrubs up to 1 m ($3\frac{1}{4}$ ft) high, composed of flat, oval joints 5—15 cm (2—6 in) long and coloured emerald-green. The flowers, borne at the tips of the terminal joints, are yellow and measure about 5 cm (2 in) across. The areoles do not have spines, merely tufts of glochids, which are very short, weak spines covered with barbs that are invisible to the naked eye. When touched, they come loose from the areole, catch in the skin and cause unpleasant itching. For this reason *O. microdasys* is not very popular in Mexico. When ingested by domestic animals it causes digestive disorders, and during wind storms can cause inflammation and even blindness if the glochids get into the eyes. *O. microdasys* is an undemanding plant that tolerates varied types of substrates, a degree of exposure to sunlight and intervals of watering. It can be grown well in light apartments where even in winter it requires occasional, moderate watering to keep the young, as yet immature, joints from drying out too much. The easiest method of propagation is by rooting individual joints.

2

Opuntia microdasys (1) is a variable species not only in the size of the joints but also in the colouring of the glochids. Although the ones most commonly grown are specimens with golden-yellow glochids, also attractive are those with brownish-red glochids, known as var. *rufida* (2) and others with white glochids, known as var. *albispina*. Glochids (3) are a typical feature of all members of the genus *Opuntia*. They are miniature barbed spines less than 1 mm long. These are most readily removed from the skin by dripping melted wax on the affected spot and, when the wax has hardened, pulling it off along with the glochids. Recommended to growers who wish to avoid these problems is the cultivar *O. microdasys* 'Angel's Wings' (4). Its glochids are white like those of var. *albispina* but are so fine that they will not pierce the skin.

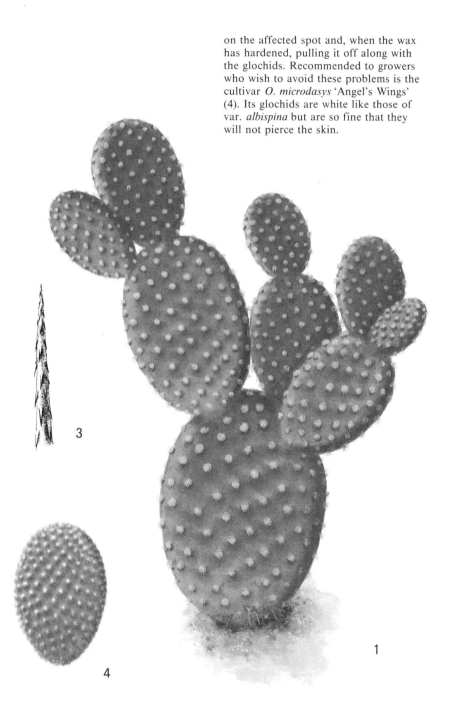

3

4

1

Pilosocereus palmeri (ROSE) BYLES ET ROWLEY

Members of the genus *Pilosocereus* belong to the group of tropical columnar cacti. They are distributed from the warm regions of Mexico, through central America, the Antilles and Bahamas to South America, where their range extends to Peru and Brazil. *P. palmeri* is one of the most attractive and rewarding cacti from the grower's viewpoint. It is found along the coast of the Gulf of Mexico in deciduous shoreline forests. Its stems penetrate the tree tops or grow on rocky outcrops devoid of taller vegetation. *P. palmeri* is 3—5 m (10—16 ft) high and about 8 cm (3 in) across and is generally much-branched in old age. From the areoles on the edges of the ribs of adult specimens there grow not only needle-like spines, but also delicate 'hairs' that merge to form a continuous white border that stands out in prominent contrast to the bluish skin of the sharp-edged, extremely narrow, protruding ribs. *P. palmeri* grows rapidly and, if provided with the right conditions, increases by as much as 15 cm (6 in) in a year. It requires a warm, moderately shaded location, a relatively nourishing substrate and liberal application of water. Because it forms an extensive root system, it should be put in a large container. Like other thermophilous cacti (e.g. melocacti), this species should also be overwintered at a higher temperature, which should not drop permanently below 12—15°C. Only thus is it possible to avoid fungal diseases of the skin, which cause unattractive spots on the top of young, as yet immature, parts of the stems.

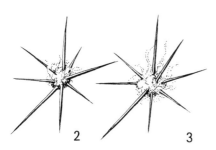

2   3

If *Pilosocereus palmeri* (1) attains a height of approximately 1 m (3¼ ft), the crown begins to be topped by an inreasing number of soft hair-like spines, jutting outwards at first but later hanging downwards, that indicate the plant has reached maturity. Then, at summer's peak, bell-like flowers, about 5 cm (2 in) across and coloured pale pink to carmine-red, emerge on the side of the stem. Cactus growers with a low glasshouse are advised to cut off the top

1

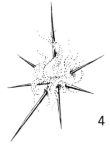

4

of the stem and put it in a pot to root.
They will thus obtain a plant with
hair-like spines adorning the edges of
the ribs along their entire length and
giving it an interesting appearance.
These plants produce flowers when
almost 50 cm (19 ½ in) high. Whereas in
younger seedlings the areoles are
practically devoid of hair-like spines (2),
older plants possess them in great
abundance (3, 4).

Selenicereus grandiflorus (L.) Britt. et Rose

The genus *Selenicereus* was named after the Greek moon goddess Selene. It includes about twenty species, all with large gleaming white flowers that have a penetrating fragrance and bloom for only one night. They are climbing plants that germinate on the ground but whose long stems clamber up trees and other suitable supports, forming huge clumps at the top that weigh tens of kilograms.

Without a doubt the most familiar of them all is *S. grandiflorus*, commonly called 'Queen of the Night'. It was described by Linnaeus as far back as 1753 and has been cultivated for a long time. It is found in the state of Veracruz, Mexico, and in Jamaica, Haiti and Cuba. It was formerly used in medical practice because the flowers and young stems contain cardiac glycosides. A synthetic substance is used nowadays in their place, and *S. grandiflorus* has ceased to be grown for this purpose. It may be used to good effect, however, as stock for grafting. Its stems, which may be up to several metres long, catch hold of the support by means of special clinging roots growing from the sides of the stems. In each areole there are seven to eleven whitish, thin, needle-like spines. The flowers emerge on the sides of the upperparts of the stems. In the final stage of their development, the growth of the buds is extremely rapid — during the last four days they may increase in length by as much as 15 cm (6 in). *S. grandiflorus* is a vigorously growing species that is good for covering the rear sections of glasshouses, verandas and the like that are not too exposed to sunlight. It requires a humus-rich, nourishing, loose substrate and, during the growing period, liberal application of water and feed. It flowers profusely after overwintering in dry conditions at a temperature of about 10°C (50°F). Propagation is by means of seeds, which germinate readily, and stem cuttings, which root rapidly.

Selenicereus grandiflorus (1) has huge white flowers, up to 25 cm (10 in) long and when fully open up to 30 cm (12 in) across, with an extraordinarily large number of stamens. The flowers appear from summer until early autumn and have a penetrating vanilla-like fragrance. They bloom very briefly. The buds open round 10 p.m., the perianth segments unfurl, after midnight they close again, and by the morning they have already wilted. In the moist and warm environment in which selenicerei grow, the flowers are soon attacked by moulds and it is for that reason that they remain open for such a short time. Larger specimens bear a profusion of flowers that open and wilt in succession. In their native habitat they are pollinated by moths.

1

Thelocactus bicolor (GAL.) BRITT. ET ROSE

Plants of the genus *Thelocactus* are rather small, flat-topped, globose to cylindrical cacti with ribs divided by deep transverse grooves into angular tubercles. The genus comprises some twenty species, distributed from southern Texas to central Mexico. They are closely related to the genus *Coryphantha*.

The body of *T. bicolor* is variable in shape as well as size. Generally it is ovoid, but in some forms or varieties it may be spherical to slightly flat-topped; in others conical to cylindrical. The average height is approximately 10—15 cm (4—6 in). The ribs are continuous or more or less divided into tubercles. Most pronounced is the variability of the spines, in terms of density and length as well as colouring. Low, flat, yellow-spined plants were described as var. *flavidispinus,* slender, white-spined specimens as var. *bolansis,* and plants with flat, flexible, straw-coloured spines as var. *schottii. T. bicolor* and other closely related species are moderately difficult to grow. If they are watered only during peak temperatures and when the substrate is absolutely dry, they will have little trouble in retaining their own roots. More difficult to grow are plants from Texas, particularly var. *flavidispinus* and var. *schottii,* which also readily lose their root system. For that reason it is recommended that these varieties, as well as several closely related species, chiefly *T. schwarzii,* be grafted on *Cereus jusbertii* stock.

Thelocactus bicolor (1) has several varieties. Var. *bolansis* (2) is readily distinguished from var. *bicolor.* Its tangled spines are pure white, and only occasionally red or yellowish as they emerge. Certain of its closely related species might also be viewed as varieties. Like *T. bicolor,* all have large violet flowers with a silky sheen and glowing red throat. They include primarily *T. wagnerianus, T. schwarzii,* and *T. heterochromus* (3), which differs markedly in habit — its body is globose, up to 15 cm (6 in) across, and the ribs are divided into large rounded tubercles.

2

3

1

Thelocactus hexaedrophorus (Lem.) Britt. et Rose

This lovely, solitary cactus was named after the shape of the mammillae of which the stem is composed (*hexaedrophorus* means bearing hexagons). It is native to the inland regions of Mexico, occurring in relative abundance in the states of San Luis Potosí, Nuevo León, Zacatecas and Tamaulipas. There, at elevations of approximately 2,000 m (6,500 ft), it grows mostly on calcareous substrates in open country devoid of taller vegetation. The body is flat-topped, globose to cylindrical and in old age nearly 15 cm (6 in) across. The surface of the skin, especially in aged specimens, is whitish-grey, as if coated with flour, giving the plant an extremely attractive appearance. The tubercles are sometimes small and not very prominent, at other time large but flatter, and in some populations nondescript and forming more or less continuous ribs. The spines are likewise variable — in shape, thickness, length, number and colouring. The flowers of cultivated cacti appear early in spring and several times more during the course of the summer. They are white, creamy-white or pale pink. They often appear on small seedlings no more than 3 cm (1 in) across, but then they are much smaller. *T. hexaedrophorus* has large seeds that germinate well, so sowing them is not difficult. Small seedlings, as well as adult plants, require moderate shade and liberal watering during the growing period. Because they have a strong turnip-like root, they should be put in a sufficiently deep container and a sandy-loamy substrate without a great concentration of organic admixtures. In winter, the temperature should be kept between 5—15°C (41—59°F).

2

Thelocactus hexaedrophorus (1) has extremely variable spines. These may be a light colour, for example yellowish-white, or carmine-red (2). In some localities plants with rather short spines, 1—2 cm ($\frac{1}{3}$—$\frac{3}{4}$ in) long, predominate, elsewhere the spines are up to about 4 cm (1$\frac{1}{2}$ in) long. One of the most distinctive geographical forms was described as var. *decipiens* (3), which has extremely short, recurved spines. There is also variability in the size of the flowers. In some specimens

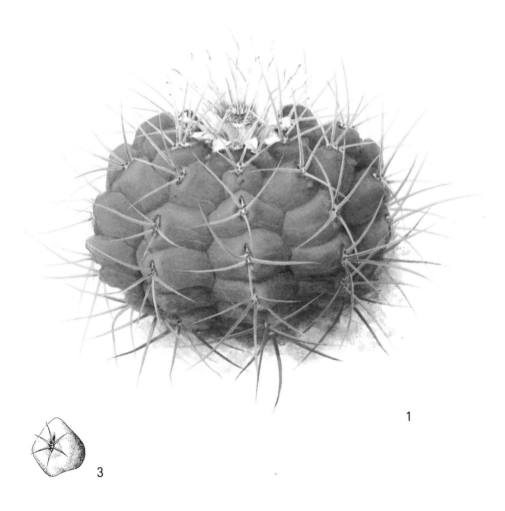

1

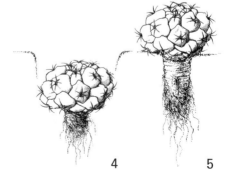

3

they are up to 8 cm (3 in) across, in
others only 4 cm ($1\frac{1}{2}$ in).
T. hexaedrophorus is protected against
the sun's rays by the white reflective
coating on the skin and also by the
plant's subterranean manner of growth
during the resting period. During the
dry winter months the plant is drawn
down into the ground (4) and not until
the onset of congenial summer weather,
during the rainy season, does its
silvery-grey body rise above
the surface (5).

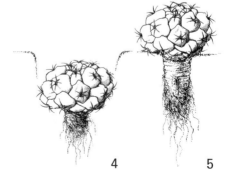

4   **5**

Turbinicarpus schmiedickeanus (Böd.) Buxb. et Backeb.

Turbinicarpus schmiedickeanus is found in the semi-desert of central Mexico, in the states of San Luis Potosí, Nuevo León and Tamaulipas, on limestone rock outcrops where it grows in full sun in crevices or pockets filled with permeable soil. Because its inconspicuously coloured dry fruits are not eaten by birds, and thus there is no means by which its seeds could be dispersed over great distances, it has only an insular distribution in small localities. It forms a number of populations that can usually be classified as distinct varieties.

Var. *schmiedickeanus* has an ovoid, grey-green body. The ribs are divided into elongated tubercles. The areoles at the tips of the tubercles bear one to four spines. These are 2.5 cm (1 in) long, flat, grey and directed upwards towards the crown. The pinkish flowers are small, only about 2.5 cm (1 in) across. They emerge only once during the season — in late winter or early spring. More rewarding are the other varieties of this species, which produce flowers almost continuously throughout the entire growing period. Examples are var. *gracilis*, which has striking, slender, glowing grey-green tubercles, var. *dickisoniae*, growing only a few kilometres away, which has numerous radiating white radial spines and one to three upward-directed dark central spines, and var. *flaviflorus*, which is characterized by pale yellow flowers. These plants are grown most reliably when grafted on *Eriocereus jusbertii* or *Echinopsis eyriesii* stock. They require a very sunny situation.

Var. *macrochele* (1) is the most wildly spined variety of *Turbinicarpus schmiedickeanus*. It has been grown for more than half a century and is often classed as a separate species. It differs from var. *schmiedickeanus* by having flatter and shorter tubercles and up to 4-cm-long ($1\frac{1}{2}$ in) intertwined spines. Its opposite is var. *klinkerianus* (2), which

2

108

1

has very few spines. From each areole
grow only three spines, twisted like
wires, two of which are not permanent
and will, after a time, come loose and
fall. The creamy-white flowers are
funnel-shaped and measure only 1.5 cm
($\frac{1}{2}$ in) across. Similar in appearance to
this variety is var. *schwarzii*; its flowers,
however, are pinkish and twice as large.

CACTI OF SOUTH AMERICA

Acanthocalycium violaceum (WERDERM.) BACKEB.

The genus *Acanthocalycium* is native to northern and north-western Argentina. It is a relatively small genus comprising only a few species. Most commonly grown is the species named *A. violaceum* after the colour of its flowers (*violaceum* meaning violet). It grows in northern Argentina in the province of Córdoba at an elevation of about 1,000 m (3,000 ft). Its stem is pale green, up to 13 cm (5 in) across and 20 cm (8 in) high, and does not produce offshoots. Growing from the areoles on the edge of the sharp-edged ribs are spines 2—3 cm (1 in) long and coloured yellow-brown. The flowers emerge on the crown several times during the growing period and remain open for two to three days. They are about 7 cm ($2\frac{3}{4}$ in) long and coloured pinkish-violet to pale violet. Like other species of this genus, this cactus is allogamous. Stiff, pointed scales that look as if they have been cut from parchment grow through the ovary and flower tube. The globose fruits contain 1-mm-long, black, finely rough-surfaced seeds. As *A. violaceum* does not produce offshoots, it is propagated only by means of seeds. The seedlings grow rapidly and their root system is resistant to fungal diseases, therefore they need not be grafted. The plant requires a sunny but moderately shaded and well-ventilated situation; in full sun with limited ventilation the skin is readily scorched. In summer, it stops growing and if watered frequently the roots will rot. In winter, it requires cool and absolutely dry conditions, otherwise it comes to life and begins to grow, thereby causing deformation of its shape in the region of the crown. Watering should be resumed in late spring.

2

Acanthocalycium violaceum (1) has a flat-topped globose body that later, but not until advanced age, will begin to lengthen slightly. The spines in the areole may be sparse or dense (2). Another, quite different species that cactus growers find rewarding is *A. glaucum* (3). In collections it may also be encountered as a variety of the species *A. thionanthum.* The distinguishing feature of *A. glaucum* is

the large flowers, up to 6 cm ($2\frac{1}{3}$ in) long, that are produced even by small seedlings measuring less than 2 cm ($\frac{3}{4}$ in) across. The skin is pale bluish-grey and the emerging spines are brown to black. The stem is ovoid in youth; in adult plants it attains a height of 15 cm (6 in) and a diameter of 7 cm ($2\frac{3}{4}$ in). It is native to the province of Catamarca in northern Argentina.

3

1

Blossfeldia liliputana WERDERM.

Blossfeldias are among the smallest of cacti, bearing flowers when they measure only 4—6 cm ($1\frac{1}{2}$—$2\frac{1}{3}$ in) across. They are distributed throughout a relatively large region in northern Argentina and Bolivia where they grow at elevations of 1,500—2,000 m (5,000—6,500 ft) on the steep sides of ravines to which they hold fast with their wiry roots. Even though they form colonies numbering thousands of individuals, they easily escape notice. They spread by means of seeds, which catch hold even on vertical rocks because their surface is covered with minute hair-like projections. The genus *Blossfeldia* was established in 1937. In ensuing years specimens were gathered in other localities and described as new species, for example *B. fechseri*, *B. minima*, etc. The differences between them are merely of a quantitative nature so that nowadays they all tend to be considered as forms of the species *B. liliputana*.

The individual body of this species measures approximately 1.5 cm ($\frac{1}{2}$ in) in diameter. During the resting period this cactus becomes greatly flattened and merges with the terrain. The small spineless areoles are white-felted and the flowers are creamy-white to yellow. In collections this clustering species is generally grown grafted on stock because growing it from seed is quite difficult, even though obtaining seeds is no problem, inasmuch as *B. liliputana* flowers readily twice to three times a year and generally produces seeds without cross-pollination. The seedlings are minute and grow very slowly during the first two years. Rooting offshoots is also difficult, for they generally dry up before they can root, and so the only reliable means of propagating this species is by grafting offshoots on stock. In summer, the plant should be kept in a moderately shaded location, and in winter, conditions should not be too cool.

With its miniature body, *Blossfeldia liliputana* (1) is most like the cacti of the genus *Frailea*, which are also among the smallest cacti of South America. Of the several dozen species, surely the loveliest is *F. asteroides* (2), sometimes also named *F. castanea*. Its body is globose, about 3 cm (1 in) across, with flat ribs and spines only a few millimetres long and pressed close to the body. The yellow flowers measure less than 4 cm (1½ in) in diameter. Unlike blossfeldias, *F. asteroides*, like other members of the genus *Frailea*, has large shell-like seeds that are often produced by self-pollination without the opening of the flower. Because it is native to the relatively moist regions of southern Brazil (Rio Grande do Sul state), small seedlings do not tolerate dry conditions in winter. Its root system is rather delicate and therefore it is recommended that it be grafted on stock.

2

1

Borzicactus aureispinus (Ritt.) Rowley

This cactus of snake-like habit with gleaming yellow spines appears in collections under several names, frequently, for example, as *Hildewinteria aureispina*, the name given to it by F. Ritter upon discovering it in the Bolivian province of Florida, where it grows on rocky terraces. Today it appears that the existence of *Hildewinteria* as a separate independent species is unwarranted.

Borzicactus aureispinus has prostrate stems up to 1.5 m (5 ft) long and about 2.5 cm (1 in) across. Whereas the skin is a vivid green, the short, dense, nonprickly spines are golden. The flowers emerge on the sides of the stems. They are allogamous, aproximately 5 cm (2 in) across, coloured orange-red and remain constantly open for about four days. *B. aureispinus* is an undemanding cactus. Its growth is rapid and it soon starts branching. In a nourishing substrate it bears flowers for the first time in its fifth to sixth year, flowering repeatedly throughout the entire growing period. If it is overwintered in a heated glasshouse or a light apartment, it may begin growth as early as February and the growing period may last until December. If space is limited, water it only occasionally. In such conditions the shoots are weaker and slighter, but are covered with dense, deep yellow spines. For profuse flowering, it should be placed in a sunny, relatively warm situation and overwintered at a temperature of about 15°C (59°F). If the temperature drops permanently below freezing point, it suffers physiological disorders that will cause spots on the skin.

2

In its native habitat *Borzicactus aureispinus* (1) has either creeping or trailing stems. This attribute may be used to good effect in the glasshouse, where the plant can be hung above ledges, or else the glowing yellow spines can be used to brighten an epiphytic arrangement in an apartment. Similar in appearance is the species *Borzicactus samaipatanus* (2), which also forms huge

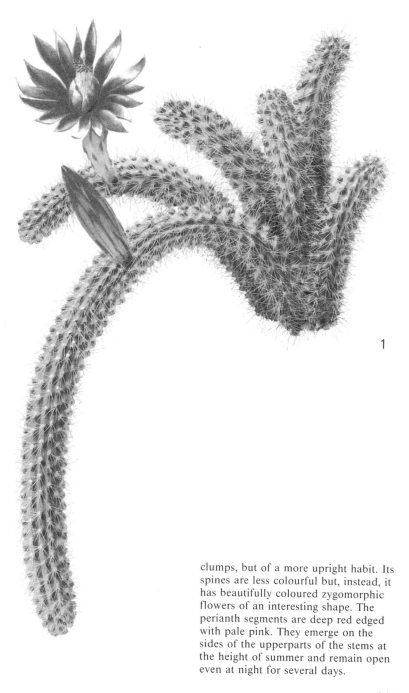

1

clumps, but of a more upright habit. Its spines are less colourful but, instead, it has beautifully coloured zygomorphic flowers of an interesting shape. The perianth segments are deep red edged with pale pink. They emerge on the sides of the upperparts of the stems at the height of summer and remain open even at night for several days.

Cereus peruvianus (L.) Mill.

Cereus peruvianus is one of the longest-known cacti. We do not know precisely where its original habitat was located, all we do know is that it is indigenous to Brazil. Like certain opuntias, it spread secondarily to other places, not only in South America. It is a columnar cactus, reaching a height of 6—7 m (20—23 ft) and branching profusely in old age. The individual branches are 30—60 cm (12—24 in) long and have a bluish bloom that later turns green. The slender ribs are very prominent and 4—6 cm ($1\frac{1}{2}$—$2\frac{1}{3}$ in) long. The felted areoles bear up to 2.5-cm-long (1 in), very prickly spines coloured brown to black. The flowers are large, up to 15 cm (6 in) long, and white; they open after sundown and bloom for just one night. Their pleasant, penetrating fragrance attracts the hawk moths that pollinate them. *C. peruvianus* is a rapidly growing, undemanding and adaptable cactus. If supplied with adequate nourishment, it will reach a height of several metres and branch profusely within twenty years, and can thus be grown only by owners of larger glasshouses. In smaller glasshouses, it is recommended that the top section of the stem be cut off and rooted. After several years, even though quite small, for instance about 20 cm (8 in) high, *C. peruvianus* will bear flowers on this cut-off top section. This cactus is so hardy that it can be grown freely outdoors from spring until autumn, for example on a patio or beside the house entrance, and in winter kept in a hallway or a room where the temperature does not drop below 5°C (41°F). It requires a well-draining, nutrient-rich substrate and adequate watering in summer. It is readily propagated by means of seeds as well as by stem cuttings and offshoots. *C. peruvianus* is a very good stock, especially for South American cacti, which retain their broad shape and dense spines when grafted on it.

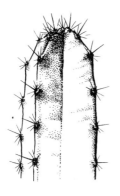

2

Cereus peruvianus (1) has a straight, unbranched stem when young. On reaching a height of about 1 m (3 ft) it begins to produce offshoots on the upperpart of the stem and turns woody at the base. The number of ribs is extremely variable; there are generally five to six but sometimes only four (2), and at other times as many as nine.

Their number usually increases with age. *C. peruvianus* has a monstrous form (3) that was described in the nineteenth century as an independent species, *C. monstrosus*. It is merely an interesting, ornamental growth deviation that is readily propagated by rooting cut-off parts of the stem.

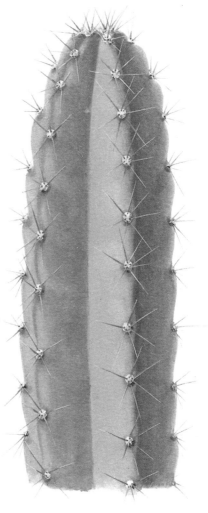

1

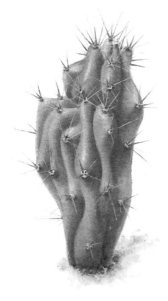

3

Chamaecereus silvestrii (Spegazz.) Britt. et Rose

Translated, *Chamaecereus* means small or minute cereus. The individual stems of *C. silvestrii*, the only species in this genus, truly resemble miniature cerei. In terms of kinship, however, *C. silvestrii* comes closer to the genus *Rebutia* or *Lobivia* and is even considered to be a *Rebutia* or *Lobivia* by some botanists. Formerly it was purposely crossed, for example with members of the genus *Lobivia*, in order to obtain larger or differently coloured flowers. This produced hybrids with thicker stems and more profuse flowers coloured orange, violet or yellow. In the wild, *C. silvestrii* is found in the mountain regions of northern Argentina, where it grows on stony terrain, and also in places moderately covered with scrub, between the towns of Salta and Tucumán. The small stems produce numerous offshoots and form huge clumps. The stems are 7—14 mm ($\frac{1}{4}$—$\frac{1}{2}$ in) across and up to 10 cm (4 in) long. The white spines are bristle-like, soft and only 1—1.5 mm ($\frac{1}{20}$ in) long. The funnel-shaped flowers, about 4 cm ($1\frac{1}{2}$ in) across, emerge in spring and often repeatedly throughout the summer. They are coloured orange to scarlet. *C. silvestrii* is easy to grow. Because it produces numerous offshoots it is mostly propagated by vegetative means. It grows extremely well even in a window garden and in summer may be put out on a patio or in the garden. Overwintering is the same as for other mountain cacti — in cool conditions and without water. To achieve profuse flowering, not only should it be provided with sufficient nutrients, especially phosphorus, but the stems should be allowed to dry up partially before the onset of the resting period, in which condition it will even withstand temperatures below freezing point.

2

Chamaecereus silvestrii (1) has a pale green skin. When exposed to intense sunlight, however, the skin partially reddens (2). A mutation without chlorophyll, called f. *aurea* (3) after the golden-yellow colour of the stems, is often propagated in horticultural establishments. Because it is incapable of photosynthesis it can be grown only as a graft. In summer it requires a light but not too sunny position, and in winter preferably warmer conditions and occasional moderate watering of the stock.

3

1

Cleistocactus strausii (HEESE) BACKEB.

Cleistocacti are slender, cylindrical, very hardy cacti. The genus is not very large, comprising approximately 40 species. Some have creeping stems, others have upright stems. One species with an upright columnar stem is *C. strausii,* considered to be the most attractive of them all because of its dense, silvery-white spines and red flowers. It grows in the mountains of the Bolivian province of Tarija at an elevation of about 1,750 m (5,100 ft). At first it remains solitary, later it produces offshoots at the base of the stem and forms rather small groups. It reaches a height of about 1.5 m (5 ft) and a breadth of approximately 8 cm (3 in). The skin is covered with straight white spines that become darker on the lower part of the stems of older plants or on stems that are no longer growing. The autogamous, tubular, red flowers emerge on the sides of the stems near the crown and are cleistogamic, which means that the floral envelopes of the hermaphrodite flowers do not open when the stamens and pistils are ripe and are thus pollinated by their own pollen. Flowers are produced by specimens 50 cm ($19\frac{1}{2}$ in) high or higher. *C. strausii* is a hardy species that may be put out in the garden in summer. It grows best in a glasshouse where, in a sufficiently nourishing and free-draining substrate, adult plants have a growth rate of up to 20 cm (8 in) per year. During the growing period it requires regular watering and only a moderately shaded, adequately ventilated situation. Best in winter are dry conditions and a temperature between 5 and 15°C (41—59°F). Propagation is by means of seeds and offshoots.

2

Cleistocactus strausii (1) is a slender, columnar cactus with lovely white spines. These are not very dense in small seedlings; only in older plants does the stem become very much thicker with more and denser spines. Similar in habit and covering of spines, but not related to *Cleistocactus,* are the members of the genus *Haageocereus.* This genus comprises some 50 species, all with thick yellow spines. They are native to the coastal and higher semi-desert regions of Peru. Their stems are only a few centimetres thick, but up to 3 m (10 ft) long, partially erect or creeping. In

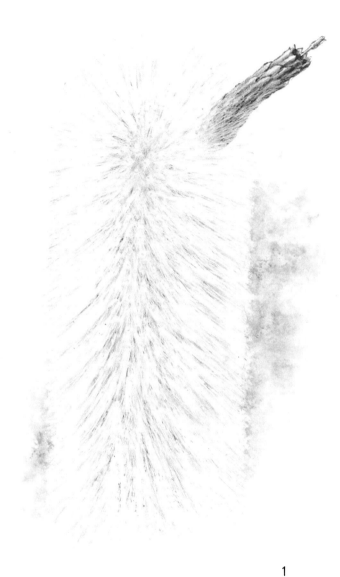

1

collections most of them are extremely
attractive plants but older specimens
must be tied to a support. This also
holds true for the most commonly
grown species *H. versicolor* (2), which, at
maturity, may be 1.5 m (5 ft) high and
only 5 cm (2 in) across.

Copiapoa cinerea (PHIL.) BRITT. ET ROSE

The genus *Copiapoa* comprises several dozen species of which the most attractive are, without doubt, the so-called white species. These are plants whose stems are coated with a layer of crystalline wax that gives them a greyish-white colour. This wax layer is particularly pronounced on plants in their native habitat, where it protects the plant's skin against the effects of strong sunlight and at the same time prevents the skin from becoming unduly damp in dense mists. These cacti are distributed in a narrow belt along the coast of northern Chile where there is practically no rainfall the whole year long, so they are therefore dependent for water on the dense mists that roll in from the ocean.

In these inhospitable regions *C. cinerea* grows on rocky plateaux and hilltops in pure mineral soils without any companion vegetation. The solitary or moderately clustering stems measure about 20 cm (8 in) in diameter and more than 1 m (3 ft) in height in adult specimens. The black spines stand out in sharp contrast against the greyish-white skin, from which the plant takes its name (*cinereus* means ash-grey). The flowers are nondescript, pale yellow, about 3 cm (1 in) across and grow from the greyish felt on the crown. Growing the plants on their own roots is not at all simple. To obtain successful results as well as more rapid growth, it is best to graft them on any of the commonly used rootstock. If you want to grow this species on its own roots then provide it with a pure mineral, well-draining substrate and apply water only lightly and occasionally. The formation of the waxy layer may be promoted by putting the plants in a warm, sunny situation with greater atmospheric moisture. At the beginning of the growing period they require shade and in winter cool conditions.

Cultivated specimens of *Copiapoa cinerea* (1) have only a thin waxy layer compared to that of imported plants. The covering of spines, however, usually remains constant, even though there are marked differences between the various forms. Of interest, for example, are specimens with two central, and no radial, spines (2). A number of plants are closely related to *C. cinerea*; these are classed either as independent species or as varieties of *C. cinerea*, the latter

3

being more correct from the botanical viewpoint. These include var. *albispina*, var. *columna-alba*, var. *dealbata*, and the one most commonly found in collections, var. *haseltoniana* (3). The latter variety is readily distinguished from var. *cinerea* by its more slender, orange-yellow spines.

2

1

Copiapoa krainziana Ritt.

In every genus of the cactus family one will probably find at least one species that differs markedly from the rest. In the genus *Copiapoa* it is *C. krainziana*, which has a thick cover of slender flexible spines. Without knowledge of its habitat and without taking its flowers into consideration, one might think it belongs to one of the groups of North American cacti. However, like all the other members of the genus, it is native to Chile, where it grows in the mountains north of the town of Taltal. In old age it forms clumps up to 1 m (3 ft) across, composed of dozens of heads about 10 cm (4 in) in diameter. A single areole may bear some 40 spines measuring up to 4 cm ($1\frac{1}{2}$ in) in length; the central spines are practically indistinguishable from the radial ones. The flowers are pale yellow, 3.5 cm ($1\frac{1}{3}$ in) across, sessile and readily escape notice in the tangle of spines on the crown. *C. krainziana* does not bear flowers readily in cultivation; this may be partly promoted by removing offshoots. In its native habitat *C. krainziana* grows in extremely dry regions in stony, mineral soils. This is also the reason why it does not tolerate organic admixtures in the growing medium nor frequent or liberal watering. To obtain successful results as well as to prevent rotting of the roots, it is best to graft it on a commonly used rootstock such as *Eriocereus jusbertii*. Like other members of this genus, it should be provided with rather cool conditions in winter (a temperature of less than 10°C/50°F) thereby preventing undesirable growth and unattractive deformations in the crown.

2

The loveliest specimens of *Copiapoa krainziana* (1) have pure white, very long, almost hair-like spines that completely cover the plant. Besides these there are also dark-spined forms with spines partly tinged brown (2) or black. The form with rigid, rather short and not too many spines was described as *C. krainziana* var. *scopulina* (3). This plant shows signs of kinship with the species *C. cinerea*, or more precisely with some of its varieties, for example var. *haseltoniana* or var. *albispina*. It also occurs, and even interbreeds, with them in some places.

1

3

Echinopsis eyriesii (Turpin) Zucc.

Echinopsis eyriesii is perhaps the most familiar and most common cactus in collections. It is also often grown as a houseplant for room decoration. Because of its undemanding requirements and excellent growth properties, it is also tried-and-tested stock for grafting. The first specimen was discovered in Brazil, and it was later also found in Uruguay and Argentina. Nowadays, known finds are only from Uruguay, where *E. eyriesii* grows together with grasses and other cacti. In youth its body is globose, in old age it is greatly elongated and up to 1.5 m (5 ft) high, with eleven to sixteen prominent, straight ribs. In young specimens the areoles are felted; later, however, they are practically bare. The spines are short, rigid, and awl-shaped. The pure white, fragrant flowers, growing from older areoles near the top, have a long tube covered with scales and grey hairs. They are up to 25 cm (10 in) long and on older plants several may open at the same time. They open towards evening and generally wilt the following day; only in cooler weather do they last for two or three days. *E. eyriesii* has no special requirements. In summer it may be grown freely on the windowsill or in the garden. To obtain greater growth rate and profuse, repeated flowering, it is often moved to a nourishing substrate with sufficient organic admixtures, and watered liberally. In winter, it requires absolutely dry conditions and a temperature of 5—15°C (41—59°F). Propagation is by means of offshoots as well as seeds.

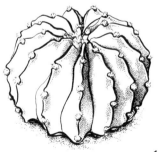

1

Echinopsis eyriesii (1) has very short, rigid spines and pure white, pleasantly scented flowers. Grown more often, however, are its hybrids. *E. eyriesii* was cross-bred with closely related species both unintentionally, because of the readily mistaken identity of the species, as well as purposely to obtain more colourful flowers and more vigorous and earlier-flowering specimens. Generally *E. eyriesii* was crossed with the white-flowering species *E. turbinata* and *E. tubiflora* or with the pink-flowering but odourless species *E. oxygona* (2).

Plants with rather long, jutting spines
and white, odourless flowers (3) point to
the hybridization of *E. eyriesii* and
E. oxygona. Note should also be taken
of the so-called 'paramont' hybrids, with
brightly coloured flowers. These were
produced by crossing *E. eyriesii* with
certain species of *Echinopsis* and *Lobivia*
that bear coloured flowers.

2

3

127

Echinopsis kermesina (Krainz) Friedrich

This red-flowering species from northern Argentina is often found in collections under the generic name *Pseudolobivia*. The person mainly responsible for the spread of the species under this name was C. Backeberg, who attempted to prove that *Pseudolobivia* and *Echinopsis* were separate, independent genera. He justified the transfer of certain species of *Echinopsis* to the genus *Pseudolobivia* on the basis of a certain dissimilarity in the flowers as well as differences in the shape of the ribs and stems. According to the latest opinions, however, botanists tend to refute *Pseudolobivia* as an independent genus. Some are also of the opinion that *Echinopsis kermesina* is merely a variety of *E. mamillosa*. *E. kermesina* has a globose body that later becomes slightly elongated and that measures about 15 cm (6 in) in diameter. It is coloured deep green and the ribs are moderately tuberculate. The radial spines are up to 1.2 cm ($\frac{1}{2}$ in) long and the central spines up to 2.5 cm (1 in) long. They are needle-like and coloured rusty-yellow with dark brown tips, becoming greyish with age. The carmine-red flowers are up to 18 cm (7 in) long and 9 cm ($3\frac{1}{2}$ in) across; they are odourless. Of all the globose cacti *E. kermesina* has the most rapid growth and therefore need not be grafted. It requires a nourishing, humus-rich substrate and liberal watering every now and then during the growing period. It need not be kept under glass but may be put outdoors during this period. Because it is a mountain species, it does not tolerate the conditions in a poorly ventilated glasshouse. In damp, overheated premises rusty spots that later turn black appear on the skin. This fungal disease occurs more often in specimens that have grown too rapidly and whose tissues are not yet fully developed.

2

Echinopsis kermesina (1) has unusually large flowers coloured pale carmine-pink to deep red. The colour fades somewhat even during flowering, which usually lasts no longer than three days. The buds develop very rapidly — practically before one's very eyes. In the final phase the bud lengthens by about 1.5 cm ($\frac{1}{2}$ in) a day. When it reaches a length of 18 cm (7 in) it opens slightly towards evening, and then fully the following morning. *E. aurea* (2) is

another cactus that Backeberg classed in
the genus *Pseudolobivia*. Today some
systematists consider it to be a member
of the genus *Lobivia*. This, too, is
a mountain species native to northern
Argentina. It is easy to grow and
produces flowers readily throughout the
entire growing period. As its name
indicates, the flowers are golden-yellow,
7—9 cm ($2\frac{3}{4}$—$3\frac{1}{2}$ in) long and odourless.

1

Eriosyce ceratistes (Otto) Britt. et Rose

Eriosyce ceratistes belongs to the small genus *Eriosyce*, established in 1872, which includes several other closely related species that are much alike and often difficult to distinguish from one another. They are large, globose cacti, native to the semi-desert regions of Chile, with a discontinuous, insular distribution in a narrow belt, about 1,000 km (600 mi) long, between the towns of Santiago and Antofagasta. They occur at elevations of 2,000—2,300 m (6,500—7,500 ft) on poor, sandy-stony soils devoid of taller vegetation. Adult forms of *E. ceratistes* are huge cacti weighing many kilograms. The largest specimens ever found measured more than 50 cm (19½ in) across and 1 m (3 ft) in height. The body is spherical at first, later barrel-shaped, with 20—30 ribs, and coloured dull green. On the edges of the ribs are large yellowish-white to brown woolly areoles bearing strong spines up to 3 cm (1 in) long. The flowers are not particularly striking and are about 3.5 cm (1 ⅓ in) across and coloured orange-yellow to red. They are not often seen in collections because the plant does not flower until it measures 12 cm (5 in) in diameter. The propagation of eriosyces is rather difficult. They do not produce offshoots and even propagation from seed is not very successful. The large seeds, mostly imported, germinate poorly and the seedlings are prone to damping-off. For this reason, it is recommended that they be grafted. Whereas the soil in which they are grown must meet the requirements of the root stock (nourishing, more humus-rich soil) the conditions of their location should meet the needs of the grafted specimen. This applies to the grafted specimens of all cacti. *E. ceratistes* will produce the loveliest spines if put in a warm, sunny spot, well ventilated at night. The winter temperature should be between 5 and 12°C (41—54°F).

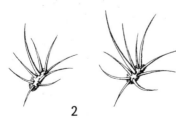

2

The spines of *Eriosyce ceratistes* (1) range in colour from pitch black through brown to amber-yellow. The spines (2) are also very variable in number and length as well as strength. Plants with yellow spines were described by C. Backeberg as *E. aurata* (3), which means golden. Correctly speaking, this is a variety or form of the species *E. ceratistes*. Eriosyces are sometimes

called the ferocacti of South America. The features they have in common with these largest of globose cacti are their huge size, flowers emerging from the latest areoles on the crown, and stout, wild spines. However, these two genera are merely similar, they are not related.

3

1

Espostoa nana Ritt.

Along with the species *E. haagei* (syn. *E. melanostele*), *Espostoa nana* is sometimes classed in the genus *Pseudoespostoa*, established in 1934 by C. Backeberg. Most botanists do not acknowledge this genus and class the two species in the genus *Espostoa*. Some ten species have been described to date but more may yet be discovered in northern Peru or Ecuador, which is the region of their distribution. In collections they are among the most attractive of the columnar cacti because of their covering of white, delicate, spidery spines and thus they are often propagated and offered for sale by horticultural establishments. *E. nana* attains a height of approximately 1.5 m (5 ft) in old age. The individual stems are up to 8 cm (3 in) across and covered by a dense tangle of white hairs, from which, in older specimens, the central spines protrude. The white flowers emerge from the cephalium which forms on the side, are up to 5.5 cm (2 in) long and open only at night. All species of this genus are native to warm regions at moderately high elevations; only very occasionally may they be found above 2,000 m (6,500 ft) in the Andes. For this reason they require a higher temperature during the growing period and a winter temperature of more than 10°C (50°F). They are grown most reliably when grafted on low, cut-back columnar stock such as *Eriocereus jusbertii* and *Cereus peruvianus*. When growing the plants on their own roots, they should be provided with a mineral substrate and watered only when it is absolutely dry and in more constant weather.

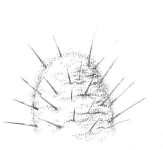

2

Espostoa nana (1) is a columnar cactus completely covered with white, hair-like spines. Similar and closely related is the species *E. haagei* (2). Whereas in youth it has only hair-like radial spines, adult forms also have large yellow central spines up to 4 cm (1½ in) long. Plants with red central spines are named var. *rubrispina*. Quite similar to espostoas in appearance is the equally attractive species *Austrocephalocereus dybowskii* (3), which has white hairs. It belongs to a relatively small genus from north-eastern Brazil. In old age it forms large clumps of slender branches that are only 8 cm (3 in) across but up to 4 m (13 ft) long and completely covered by

132

delicate spidery hairs from which protrude yellow-brown central spines measuring up to 3 cm (1 in) in length. The white flowers emerge from the dense felt of the cephalium, which is up to 60 cm (24 in) long and rises vertically from the crown.

3

1

Gymnocalycium baldianum (Spegazz.) Spegazz.

The genus *Gymnocalycium* comprises some 100 species. Most of them are found in Argentina, but they are also distributed in Bolivia, Paraguay, Uruguay and southern Brazil. Whereas some species grow in flat country at low elevations, others are found up to elevations of 3,500 m (11,500 ft). The stems are mostly globose and the flowers white to pink, only very occasionally red or yellow.

One of the most widely cultivated red-flowering species is *G. baldianum*, described by Dr Spegazzini as far back as 1905. It is found in the mountains of northern Argentina, in the province of Catamarca, at elevations of approximately 2,000 m (6,500 ft), where it grows in dry, stony localities without much taller vegetation. *G. baldianum* is a small, globose cactus, only 4—7 cm ($1\frac{1}{2}$—$2\frac{3}{4}$ in) in diameter, coloured dark grey-green. The nine to eleven ribs are divided into broad, rounded tubercles separated by relatively deep longitudinal grooves. The areoles bear three to seven spines, straight to slightly curved and 7—12 mm long. There are no central spines. The first flowers appear on seedlings in the third to fourth year. They have a long tube covered with bare scales. The flowers of various specimens of the species differ in colour. They may be carmine to deep red as well as violet-tinged to rosy-purple. *G. baldianum* is not difficult to grow. Beginners should have no trouble propagating it from seed because the seeds are large and germinate readily, and the seedlings grow rapidly from the very start. The plants also flower early and readily and are fairly hardy in cultivation.

2

Gymnocalycium baldianum (1) flowers repeatedly throughout the entire growing period. The development of the bud (2) takes about five weeks; it is protected by bare scales, from which the genus *Gymnocalycium* also takes its name. Closely related species include *G. uebelmannianum* with creamy-white flowers and the miniature, clustering *G. andreae* (3) with bright yellow flowers. These flowers are produced by two- to three-year-old seedlings in early spring and measure up to 4.5 cm ($1\frac{3}{4}$ in) in diameter. Because *G. andreae* is native

134

to the Sierra de Córdoba mountains, where it occurs at higher elevations (1,500—2,000 m/4,900—6,500 ft), it stands up well to cooler conditions in winter. However, it is prone to sun scorch particularly at the beginning of the growing period.

3

1

Gymnocalycium cardenasianum RITT.

One of the loveliest species in the genus *Gymnocalycium* is *G. cardenasianum*, a relatively recent species described by F. Ritter in 1964. It is native to Bolivia, to Mendez Province, Tarija Department. It differs from the related *G. spegazzinii*, discovered as early as the beginning of the twentieth century, primarily by its dull grey-green colour and large size. In maturity it measures up to 30 cm (12 in) across and 20 cm (8 in) in height, which ranks it among the largest species of the genus. The ribs are markedly rounded, flat, and 2.5 cm (1 in) wide. The stem is covered by large spines up to 8 cm (3 in) long. Although the original description refers only to dark spines, in collections one will generally find specimens with light-coloured spines. As in the other species of *Gymnocalycium*, the spectrum of colours in spines is very wide, ranging from yellowish-white, pink, amber-yellow, through various shades of ochre and brown to pitch black. The flowers are not very large in proportion to the size of the plant (about 5 cm/2 in across) but are very attractive in their silky sheen and pale pink colour. *G. cardenasianum* is moderately difficult to grow. It attains flower-bearing size only slowly and it is therefore recommended that it be grafted in order to speed up the process. The seedlings require a well-draining, relatively sandy substrate and watering when the thermometer registers higher temperatures.

Although *Gymnocalycium cardenasianum* (1) does not flower until a more advanced age, it then bears flowers readily from spring until winter. Following pollination, it produces 2-cm-long ($\frac{3}{4}$ in) fruits with a bluish bloom (2) that take relatively long to ripen in comparison with the fruits of the other species (generally about three months). A single fruit contains 200—250 minute, reddish seeds. The closely related species *G. bayrianum* (3) is native to Tucumán Province in northern Argentina. It is coloured blue-green to green and its ribs are very flat. The areoles bear three spines at first, later as many as five. When the

2

spines emerge they are light brown with a dark tip; later, however, they turn grey. The flowers emerge on seedlings when they attain a diameter of 4 cm (1 $\frac{1}{2}$ in). They are about 6 cm (2 $\frac{1}{3}$ in) long and 4 cm (1 $\frac{1}{2}$ in) across, and coloured cream-white with a silky sheen and reddish throat.

3

1

Gymnocalycium denudatum (LINK ET OTTO) PFEIFF.

Gymnocalycium denudatum is the type species of the genus. The first shipments of living plants reached Europe as far back as 1825. It is interesting to note that these specimens all had six to eight ribs; the original description makes no mention of five-ribbed specimens, which nowadays predominate in collections. Later several other forms were found in a number of places in southern Brazil, in the state of Rio Grande do Sul and in the border territory of Brazil and Uruguay, where *G. denudatum* grows on stony, grass-covered hillocks. The stem is globose, about 8 cm (3 in) across, with rounded ribs divided into tubercles. Each areole bears five to eight radiating spines, almost appressed to the body, and coloured yellow-brown, later grey. The flowers arise near the crown, are 5—6 cm (2—2⅓ in) long, and have a fleshy tube and, usually, white perianth segments. In specimens from some localities, however, the flowers are coloured pale to deep pink. *G. denudatum* has large seeds (up to 2 mm long), that are among the largest in the genus. When sowing the seeds, press them slightly into the substrate so that they will swell more readily. Because the conditions of its native habitat are relatively humid *G. denudatum* does not stand up well to lengthier drying out of the substrate. In the case of small seedlings, it is of particular importance that the substrate is not allowed to dry out permanently during the growing period and that they be watered once or twice during the winter. In its native habitat *G. denudatum* grows in the shade of grasses and for that reason should be provided with moderate shade. In winter the temperature should not drop permanently below 8—10°C (46—50°F).

2

Whereas in some forms of *Gymnocalycium denudatum* (1) the spines curve towards the stem only slightly, in other forms (2) they are pressed to the stem along their entire length, looking at first glance much like the legs of a spider spread out from the areole in the centre. Two related species are native to the grassy, slightly undulating plains of Uruguay: *G. netrelianum* with yellow flowers and a greater number of ribs (about 14) and *G. uruguayense*. The latter has twelve to fourteen ribs divided into prominent tubercles with wart-like processes. There are usually three (only very occasionally five) spines, measuring up to 2 cm ($\frac{3}{4}$ in) in length, in each areole. The flowers are yellow. Specimens in some localities have pale pink to pinkish-violet flowers; these are named *G. uruguayense* var. *roseiflorum* (3).

3

1

Gymnocalycium gibbosum (Haw.) Pfeiff.

On the South American continent the further south one goes in the direction of Tierra del Fuego, the fewer plant species one finds as the temperature decreases. This also holds true for the Cactaceae family. In the harsh environment of Patagonia, however, one does find several genera of cacti, with *G. gibbosum* marking the southern limit of the genus *Gymnocalycium* as well as all other globose cacti. It is the oldest known species of the genus, described as far back as 1812, when it was given the name *Cactus gibbosus*. In mature age *G. gibbosum* measures up to 15 cm (6 in) across and 60 cm (24 in) in height. It has a thick, grey-green skin, a bare crown, and twelve to sixteen blunt, prominently tuberculate ribs. It is from these that the plant takes its name (*gibbosus* means tuberculate). The spines exhibit marked variability in number, length and strength, as well as coloration. Even in a single locality one will find specimens with straight, strong, rigid spines as well as ones with weaker, flexible, twisted spines. They are generally a pale colour with a dark brown to black base, but in some plants they are darker. With age, all spines become greyish. The flowers are about 6.5 cm ($2\frac{1}{2}$ in) long, white or pink-tinged, and sometimes autogamous. *G. gibbosum* is one of the hardiest of all cacti. During the growing period it may be placed in the open, unprotected by glass, in a spot facing slightly away from the sun, otherwise it suffers sun scorch. In the glasshouse it also requires a shaded situation and adequate ventilation. In a completely dry substrate it tolerates even lengthier periods at temperatures below freezing point in winter. When watered liberally its growth is very rapid and it produces flowers for the first time in its fourth year.

2

Gymnocalycium gibbosum (1) is an extremely variable species that has many varieties. Var. *nigrum* (2) is readily distinguished from the type variety *gibbosum* by its dark skin and shorter, almost black spines. It is interesting to note that even offspring grown from imported seeds will exhibit marked variability in the colouring of the skin as well as the spines. Hence the smooth transition from var. *gibbosum* to var. *nigrum* does not bear truly convincing testimony in favour of acknowledging it as a separate, independent variety. The same may be said of var. *nobile* (3), which has longer, flexible radial spines and strikingly light-coloured central spines.

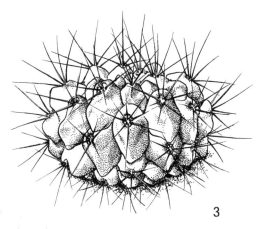

3

1

Gymnocalycium lafaldense Vaupel.

This miniature cactus is found in collections under the name of *Gymnocalycium lafaldense* or *G. bruchii*. Because there are several discrepancies in the description of *G. bruchii*, which indicate that it may not even belong to the genus *Gymnocalycium*, the author believes *G. lafaldense* to be the more correct designation. The species was described under this name in 1924. It was named after the place of its occurrence, Mount La Falda in Sierra de Córdoba in northern Argentina. There it grows up to elevations of 3,000 m (9,800 ft) on dry, sandy-stony soils as well as in rock crevices fully exposed to the sun. *G. lafaldense* has a small, dark green stem, only 3—5 cm (1—2 in) across, growing from a stout turnip-like root. It produces offshoots prolifically and forms huge clumps. Each areole generally bears twelve to fourteen narrow radial spines only about 6 mm long. The flowers are approximately 3 cm (1 in) long and 3.5 cm ($1\frac{1}{3}$ in) across and open in the afternoon over a period of three to four days. The fruits are not very large, but contain a small number of rather large spherical seeds. *G. lafaldense* is easy to grow and is generally propagated by means of offshoots. Propagation from seed is more difficult. The plant flowers in early spring when the pollen ripens poorly; even when the flowers are pollinated, the fruits contain few seeds. *G. lafaldense* stands up well to low winter temperatures and even temperatures below freezing point if the substrate is absolutely dry.

Even though the species *Gymnocalycium lafaldense* (1) has mostly violet-pink flowers, as stated in the original description, the flowers may also be nearly white or a deep pinkish-violet (2). The spines also exhibit a wide range of colour, from white through yellowish or pinkish to brown and red. In addition to this, *G. lafaldense* is also very variable in terms of the number and length of the spines, the size of the stem and the shape of the flowers. The flowers may be short and flat as well as markedly elongated. Because *G. lafaldense* is a very variable species, several forms

2

were described, all from the same locality. From a strictly botanical viewpoint, these forms are of no importance. However, the species *G. albispinum* (3) might be considered to be a variety of *G. lafaldense.* It grows in an isolated locality in the neighbourhood of Alta Gracia and differs by having a greater number of white spines.

3

1

G. mihanovichii was discovered by Czech traveller and cactus grower A. V. Frič at the beginning of the twentieth century in Gran Chaco, Paraguay, where it grows in almost flat, scrub country on almost non-porous soil. During the growing period it is shaded by shrubs; only in winter, when the region is without water for seven months and plants shed their leaves, is it exposed to the sun. At this time protection is provided by a change of its skin colour to brown. *G. mihanovichii* is a miniature cactus only 2—3 cm ($\frac{3}{4}$—1 in) high and 4—6 cm ($1\frac{1}{2}$—$2\frac{1}{3}$ in) in diameter, with a grey-green, brownish or reddish skin. The ribs are broadly triangular. There are generally five to six radial spines in each areole, although some of them later fall. The flowers are up to 5 cm (2 in) long, narrowly funnel-shaped, and coloured yellow-green to brownish-green. The seeds of this, as well as other closely related, species germinate relatively poorly in the first year after they are harvested, but their powers of germination generally improve during the second to third year. If the seeds do not germinate, take them up with forceps, dry them and sow them again the following year. Even seeds that germinate well require a lengthier period and higher temperature for germination. All plants of the *G. mihanovichii* group require a permanently shaded and very warm situation. It is recommended that buds or fruits that the plant sets on in the autumn be removed before the onset of winter, for they will readily become mouldy and may cause the plant to become infected.

2

Gymnocalycium mihanovichii (1) has several varieties, the loveliest being var. *friedrichii* (2). This has a darker skin, higher and more slender ribs, longer spines that readily fall, and pale to deep pink flowers that open out more widely. The bright red fruits contain up to 700 spherical seeds. *G. mihanovichii* develops a greater number of carotenes in the subcutaneous and photosynthesizing cells within the stem tissues, their increased concentration being the plant's means of protection against damage by strong sunlight. Japanese growers

1

3

recognized this ability on the part of
G. mihanovichii and, by selective
breeding, obtained plants with stems
coloured blood-red, orange, pink, yellow
and violet (3). Because these specimens
are incapable of photosynthesis (they
are mutants devoid of chlorophyll), they
can survive only if grafted on stock.
During the growing period they require
a warm, partly shaded situation and
regular watering; during the resting
period they like occasional watering and
a light location that is not too cold.

Gymnocalycium monvillei (LEM.) BRITT. ET ROSE

This attractive and easily grown cactus was described as far back as 1838. At that time, however, it was not the custom to concern oneself with the variability of a species, which caused marked problems in classifying plants that were imported at a later date and belonged to the same species but differed in habit. That is why *G. monvillei* is also found in collections under the name *G. multiflorum*. Like most species of this genus, it is native to the province of Córdoba in northern Argentina. Its stem is globose, up to 20 cm (8 in) in diameter, and the ribs are divided into large tubercles. A characteristic feature is the vivid green colour of the skin. The original description states that there are twelve to thirteen radial spines and one central spine in each areole, but that the latter may be absent. *G. monvillei* is a variable plant and it is not at all uncommon for it to have far fewer spines. The spines are up to 4 cm ($1\frac{1}{2}$ in) long and are generally coloured pale yellow with a purplish-red base. In some specimens the spines are nearly white, while in others they are dark. The flowers are relatively large, up to 8 cm (3 in) long, and pure white or pinkish. The most closely related species are *G. schuetzianum, G. achirasense,* and *G. horridispinum* with flowers coloured pale to deep pink. The advantage of members of the genus *Gymnocalycium,* and hence also of *G. monvillei,* is their less demanding light requirements in terms of both duration and intensity. Whereas most cacti have fewer spines and grow taller if provided with insufficient sunlight, *G. monvillei* grows well even in situations that lack sunlight all day long, for example in an east- or west-facing window garden.

3

Gymnocalycium monvillei (1) is variable mainly in terms of its spines. Whereas in some specimens each areole bears only five to seven radial spines and no central spine, in others there may be twice as many radial spines. Also frequently encountered are specimens with a larger central spine placed at right angles to the body (2). The flowers of this, and closely related, species are among the loveliest in the whole genus. They resemble water-lily blossoms in both colour and the arrangement of the

2

perianth segments. Whereas in some specimens they are pure white, in others they are tinted pink (3). They are usually smaller on young plants than on older specimens. The fruits (4) are the same colour as the plant's skin. They are up to 4 cm ($1\frac{1}{2}$ in) across and contain as many as 1,500 small seeds.

4

1

G. *saglionis* has been grown in collections for nearly one and a half centuries. In terms of horizontal as well as vertical distribution, it heads the list of species belonging to this genus. The scattered localities of its discontinuous range are in the rocky mountain sections of the north Argentinian Cordilleras, between the towns of Salta in the north and San Luis in the south, at elevations of 1,000—3,000 m (3,300—9,800 ft). The north-south axis of its range is about 1,000 km (600 mi) long. *G. saglionis* is the largest member of the genus; its globose stem may attain a diameter of up to 50 cm ($19\frac{1}{2}$ in). The ribs are divided into rounded, five- to six-sided tubercles up to 3.5 cm ($1\frac{1}{3}$ in) wide. Specimens from some localities have few spines (about eight to each areole); specimens from other localities have so many that the flowers have great difficulty in pushing through and opening. *G. saglionis* is readily distinguished from the the other members of the genus not only by its general habit, but also by its reproductive organs. The flowers are sessile, 3.5 cm ($1\frac{1}{3}$ in) in diameter as well as in length, broadly funnel-shaped to bell-shaped, and coloured pure white, greenish or pinkish; also occasionally encountered are flowers coloured a dark pink. *G. saglionis* is rather difficult to grow in the initial years. The small seeds germinate and give rise to small seedlings whose rate of growth is slow at first but rapid later.

2

The range of colours exhibited by the spines of *Gymnocalycium saglionis* (1) is extraordinarily large. The commonest are dark reddish-brown to black but one may also encounter pale-coloured spines. Several varieties have been described on the basis of the coloration of the spines but these are of importance only from the collector's viewpoint. Plants with pale, nearly white spines are customarily designated as var. *albispinum*, those with yellow spines (2) as var. *flavispinum*, those with pink

1

3

spines as var. *roseispinum*, those with black spines as var. *nigrispinum*, and those with red spines as var. *rubrispinum*. The fruits (3) of this species are up to 4 cm (1½ in) across and their watery pulp contains as many as 3,000 minute seeds. They should be left on the plants and gathered only after they split. Do not wait too long to clean the seeds, for the pulp either becomes mouldy or dries out and then washing the seeds becomes hard work. This applies to the seeds of all species of *Gymnocalycium*.

Gymnocalycium spegazzinii BRITT. ET ROSE

This extremely variable mountain species is native to the provinces of Salta, Tucumán and Catamarca in Argentina. The north-south axis of its range measures more than 300 km (188 mi). In earlier days, however, collectors did not scour the entire area and the specimens dispatched to collections came from only a small number of localities. This was the main reason for the former distorted view of the species' overall variability, reflected also in the establishment of the unwarranted variety *maior*. *G. spegazzinii* remains solitary and in mature age reaches a height of 10—20 cm (4—8 in). The crown is covered to a greater or lesser degree by greyish-white felt. The ribs are straight, divided by narrow grooves, and about 3 cm (1 in) wide at the base. Each areole usually bears five to seven radial spines; central spines occur only rarely. The flowers are about 7 cm (2¾ in) long and covered on the outside with fleshy scales edged pinkish-white. Except for a few species, gymnocalyciums are easy to grow. *G. spegazzinii* is no exception to this rule, except that its growth is slow. Therefore, those who wish to obtain flower-bearing specimens as quickly as possible are advised to graft them on stock. Opuntias are among the best of permanent stocks for all species of *Gymnocalycium*, such as *Opuntia ficus indica* and *O. tomentosa*. Specimens grown on their own roots should be provided with a well-aerated and nourishing substrate with an adequate content of organic particles. Repot the plants frequently, even every year, for quicker growth.

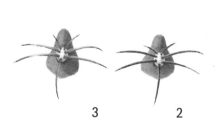

3   2

Gymnocalycium spegazzinii (1) is variable primarily in the number, length, strength, colouring, and curvature of the spines. The spines are generally brown or pitch black (2), but sometimes also red, pink (3) or white. Plants may have three, five, seven, nine or even as many as eleven spines to each areole. In length they range from 1 to 3, or very occasionally 4 cm (⅓—1½ in). There is also pronounced variability in the colour of the skin, which may be green to

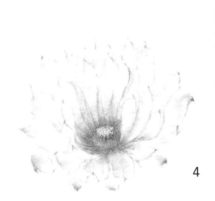

greyish-blue, or even brownish, as well as in the size, arrangement and colour of the flowers and the length of the flower tube. The flowers (4) are creamy-white to pink, less frequently brownish or greenish. The throat is always coloured pale to deep red. As in all members of the genus, the flowers open in the afternoon and bloom for about six days.

4

1

Lobivia jajoiana BACKEB.

The generic name *Lobivia* is an anagram of Bolivia, the name of the country where lobivias are most common. Over the years, however, these cacti have also been discovered in the mountain regions of neighbouring countries, namely Peru and northern Argentina. This also applies to one of the most familiar lobivias, the species *L. jajoiana*. It is distributed over a relatively large area in the northernmost regions of Argentina, where it grows at elevations of approximately 3,000 m (9,850 ft), in forms that are quite different from one another. The most pronounced of these were classified as varieties, in some cases even as separate species. The stem is globose, becoming slightly prolonged in mature age, usually about 6 cm ($2\frac{1}{3}$ in) across, with ribs divided into tubercles. It does not produce offshoots. The conspicuous central spine is 4—6 cm ($1\frac{1}{2}$—$2\frac{1}{3}$ in) long, hooked and directed upwards. The broadly funnel-shaped, massive, orange to wine-red flowers grow from areoles on the side of the stem. They have a dark purple throat and dark stamens with yellow anthers arranged in a regular circle. *L. jajoiana* is propagated only from seeds, which should be sown at a lower temperature (about 20°C/68°F). A lower temperature promotes better germination and prevents the seedlings from growing too tall, which often happens, especially if the light is rather poor. During the growing period the plants should be watered at intervals and rather liberally. Water should be withheld towards the end of summer, however, so that the plant has time to dry before the onset of winter. In the case of this cactus there is a greater risk of winter growth, which may be prevented not only by seeing to it that the substrate is thoroughly dry, but also by providing it with a cooler winter temperature of less than 12°C (54°F).

3

Like all other lobivias, *Lobivia jajoiana* (1) has very attractive flowers. They open in the morning but begin to wilt during the course of the same afternoon. Of the existing varieties perhaps the most familiar is var. *nigrostoma* (2), formerly considered to be a separate species. It has a slightly darker skin and yellow, occasionally orange or dark red, flowers with conspicuous black throats, measuring up to 8 cm (3 in) across. The loveliest is var. *paucicostata* (3), often

1

encountered in collections under the
name *L. paucicostata*, syn. *L. glauca*. Its
stem is only 3 cm (1 in) across, with
a pale greenish-grey skin and
sharp-edged ribs, and it is drawn down
almost below the level of the ground.
Another reliable means of identification
is the nearly 4-cm-long (1½ in), black
central spine. The flowers are about
5 cm (2 in) across and are pale to dark
red with a black throat.

2

Lobivia wrightiana Backeb.

The flowers of lobivias are considered to be the most attractive of the entire Cactaceae family. They are generally vividly coloured. Nevertheless, even here there are exceptions, for instance *L. wrightiana*, whose flowers, unlike those of other lobivias, are a delicate pale violet with a narrow tube and long-pointed perianth segments; unfortunately they, too, remain open for less than one day. *L. wrightiana* is a miniature species remaining solitary at first but later producing offshoots, although not until advanced age. Whereas the radial spines are only 5—7 mm long, the central spines may reach a length of 7 cm ($2\frac{3}{4}$ in) or even more. They are coloured pale to dark brown with a yellowish base. The pale violet-pink flowers, up to 5 cm (2 in) across, arise from the lower areoles, being produced by seedlings as early as the third year. *L. wrightiana* has a slow rate of growth at first, while its turnip-like root is developing, and for this reason many cactus growers choose to graft it. Recommended as stock is *Cereus peruvianus*, on which, if watered with care, it retains its natural habit and produces a lovely display of spines. It also grows reliably on its own roots. Like most cacti with a turnip-like root, it requires a heavier substrate and liberal, but not too frequent, application of water. The best location in summer is one that is sunny but slightly shaded, with plenty of fresh air, not only during the hottest period of the day but also at night. To prevent winter growth, it should be kept at a temperature of 5—12°C (41—54°F) during the winter.

2

The variable feature of *Lobivia wrightiana* (1) is its spines, which also change with the plant's age. Whereas in young seedlings the central spines are absent or short, in adult specimens they are markedly longer. Specimens with robust spines are generally grown in collections, but in their natural habitat one may encounter plants that have short spines even in old age (2). C. Backeberg described them as var. *brevispina*, i.e. short-spined. From the botanical viewpoint, however, there is

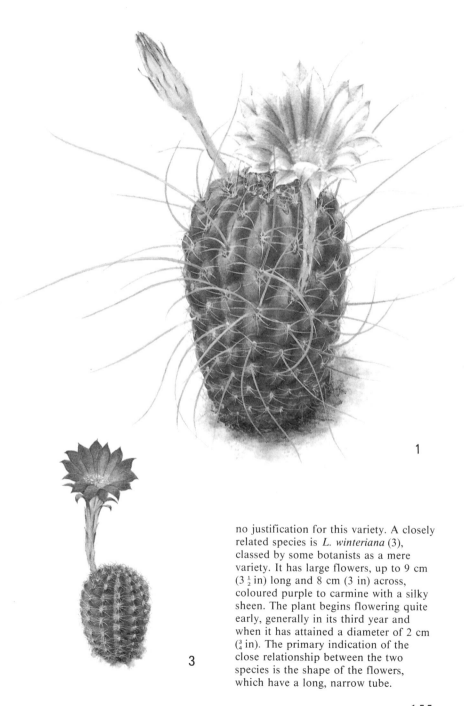

1

3

no justification for this variety. A closely
related species is *L. winteriana* (3),
classed by some botanists as a mere
variety. It has large flowers, up to 9 cm
(3 ½ in) long and 8 cm (3 in) across,
coloured purple to carmine with a silky
sheen. The plant begins flowering quite
early, generally in its third year and
when it has attained a diameter of 2 cm
(¾ in). The primary indication of the
close relationship between the two
species is the shape of the flowers,
which have a long, narrow tube.

155

Matucana aurantiaca (Vaup.) Ritt.

Matucanas are typical representatives of the mountain cacti of Peru. Some systematists do not acknowledge the independent genus *Matucana* and class its members in the genus *Borzicactus* or *Loxanthocereus*. C. Backeberg, on the other hand, divided the genus *Matucana* into two smaller genera — *Matucana* and *Submatucana*. Such a division of the genus is unwarranted from the botanical viewpoint.

M. aurantiaca flowers readily and is therefore often grown in collections. It is distributed over a relatively large area in northern Peru, at elevations of 3,000—4,000 m (9,850—13,000 ft), occurring in many, relatively different, local forms, some of which were classified as separate independent species.The stem is a glossy bright green and up to 15 cm (6 in) in breadth as well as height. It sometimes produces offshoots. The spines are narrowly needle-like and coloured yellowish-brown to reddish-brown. As in all matucanas, the zygomorphic flowers are allogamous. They are up to 9 cm ($3\frac{1}{2}$ in) long and have a massive tube.

M. aurantiaca and closely related species are easy to grow and therefore need not be grafted. During the stage of full growth, they should be placed in a sunny, partly shaded, well-aired position. At the beginning of the growing period this species is more prone to sun scorch and should be fully shaded. The illustrated species should be overwintered in an absolutely dry substrate and at a higher temperature than most matucanas (10—15°C/50—59 °F), so that the skin is not attacked by fungal diseases.

Matucana aurantiaca (1) is an extremely variable species, not only in general appearance but also in the colour of the flowers. The latter are generally reddish-yellow, but in some specimens they are nearly pure white or red. Plants with salmon-pink flowers were described as M. currundayensis, but there is no justification for this differentiation.
A closely related, yet readily distinguished, species is M. myriacantha (2). Its broadly spherical body is thickly covered with yellow or reddish-brown,

3

bristly spines. As in other matucanas the zygomorphic, readily produced flowers do not close at night. They are coloured pink to orange-red. Identical in appearance is the species *M. weberbaueri*; its flowers, however, are yellow (3).

2

1

A striking characteristic of the genus *Melocactus* is the cephalium, felted and devoid of spines, that forms on the crown, where the flowers and fruits emerge. The entire genus includes some 50 species of thermophilous cacti.

M. concinnus (meaning graceful) is one of the few melocacti that have a bluish to blue skin. It is native to the state of Bahia in Brazil, where it grows in hilly country at elevations of 1,000 m (3,300 ft) in sandy-stony soils in the shade of dwarf trees and shrubs. It was originally discovered west of the town of Seabra and was thus originally named *M. seabrasensis*. It is also found in collections under the names *M. elegans* and *M. gracilis*, which are likewise invalid. *M. concinnus* is one of the smallest of the melocacti, reaching a height of 8—9 cm $(3-3\frac{1}{2}$ in), including the cephalium, and a breadth of 10—11 cm $(4-4\frac{1}{3}$ in). The ribs are extremely sharp-edged. The cephalium is composed of dense red bristles growing from white wool. The bright violet flowers appear repeatedly throughout the summer and, as in other melocacti, are very small, less than 1 cm $(\frac{1}{3}$ in) across. Growing melocacti is moderately difficult. They grow rapidly and, by and large, reliably on their own roots. Their only requirement is warmth. In winter the temperature must not drop below 10°C (50°F) even in the case of the hardiest species, of which *M. concinnus* is one. The optimum temperature is between 15 and 20°C (59—68°F). *M. concinnus* requires a nourishing substrate and liberal watering during the growing period. It thrives in the conditions of a moderately shaded glasshouse without too much ventilation, where it is provided with water from below and by misting.

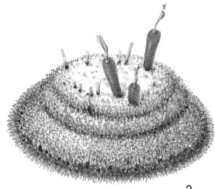

2

Melocactus concinnus (1) often forms a cephalium as early as its fifth year, when, under proper conditions, it should have reached a diameter of about 8 cm (3 in). The formation of cephalia is an indication of the attainment of sexual maturity, because it is only in the cephalium that flowers can be produced. The cephalium (2) is flat, 3 cm (1 in) high, and looks as if it has caved in. The

1

3

bright violet, fleshy fruits produced
following pollination (even
self-pollination), are more striking than
the flowers. However, they emerge from
the cephalium only during the year after
the flowers are spent. The small
seedlings (3) differ quite markedly from
flower-bearing specimens, yet their skin
is a bluish colour even at this age.

159

M. *matanzanus* in an endemic species found only on a rather small area on the northern coast of Cuba in the province of Matanzas. In its native habitat it is a rare species on the verge of extinction, however it is quite widespread in cultivation. There are several reasons for this. First, it forms a lovely cephalium with red bristles when it is only 6—8 cm ($2\frac{1}{3}$—3 in) large. Its small size makes it a suitable specimen for small collections with limited space. It also does well as a houseplant if placed in a sunny window away from draughts and protected in winter from exposure to a temperature of less than 15°C (59°C) over a longer period. M. *matanzanus* has a globose stem that becomes prolonged in mature age and measures up to 10 cm (4 in) across. It is divided into eight ribs that are relatively large at the base. The skin is bright green. The spines are curved towards the body and are less than 2 cm ($\frac{3}{4}$ in) long. The areoles are markedly woolly. All Cuban melocacti require a higher winter temperature, otherwise the colour of the skin turns pale and later the plants die. If you cannot maintain a temperature of at least 15°C (59°F) in the glasshouse, it is better to move the plants into the house. In summer, they should be placed in a warm, sunny situation but should be shaded at the beginning of the growing period. M. *matanzanus* starts forming a cephalium early, but, as with other melocacti, the process can be speeded up. When the plant measures about 6 cm ($2\frac{1}{3}$ in) in diameter, in other words attains the size when sexual maturity is close at hand, stop watering in summer and provide a temperature of up to 45°C (113°F).

2

1

The most attractive feature of *Melocactus matanzanus* (1) is the orange-red cephalium, which begins to develop in about the sixth year. Until then there is nothing particularly attractive about the plant (2), even though it differs somewhat from other melocacti. The flowers are small, approximately 5 mm across, violet and practically unnoticeable in the cephalium. They either grow singly or in the form of a wreath. As in other melocacti, they open in the afternoon for only a few hours. They are autogamous, which means that the fruits are formed even when the flowers are pollinated by their own pollen. If you wish to pollinate them yourself, carefully insert the tip of a pencil inside the flower. The fleshy, pale red fruits do not appear in the cephalium until the following year.

Neochilenia napina (PHIL.) BACKEB.

At the southern edge of Chile's Atacama Desert, in the Huasco and Copiapó river valleys, by the seacoast and then northward along the coast to the region south of the town of Taltal one may very occasionally come across sporadically distributed localities of miniature cacti. Non-flowering specimens are hard to find because their stems are almost entirely hidden in the sandy-stony soil. Other characteristic features, in addition to the miniature stems, are the ribs that are divided into small flat tubercles with submerged areoles, the short spines, and the large turnip-like root. This is a relatively specific group of cacti within the genus *Neochilenia*, which some authorities even considered to be a separate genus. Currently there are about fifteen species which can be passably distinguished from one another. Sadly, however, most are on the verge of extinction in their native land. *N. napina*, the best known representative of this group of cacti, has a globose greenish-grey body only about 5 cm (2 in) across. The ribs are divided into tubercles with areoles bearing seven to nine short black spines. The flowers are approximately 4 cm ($1\frac{1}{2}$ in) long and are coloured yellow to pinkish-red. They open in the morning over a period of about three days. The seedlings grow very slowly on their own roots because the first part to be formed is the turnip-like root and hence the growth rate of the top parts is slow. When grown in this manner *N. napina* retains not only its miniature habit but also its distinctive attractiveness. If it is provided with a mineral substrate, a position with full sunlight and minimum application of water, it may develop into a specimen that is indistinguishable from imported individuals.

3

Neochilenia napina (1) generally remains solitary and produces offshoots only if the crown is damaged, which often happens in places where domestic animals graze. In cultivation it forms offshoots primarily after being grafted, this being due to the great amount of nutrients supplied to it by the stock. Like its closely related species, *N. napina* forms a strong turnip-like root (2) whose volume is several times that of the above-ground part of the stem. Its function is to store food reserves, thereby enabling the plant to survive the extreme conditions to which it is exposed on the Chilean coast. Another typical characteristic is the reduced spines; those of *N. napina* are less than 3 mm long (3).

2

1

Neochilenia paucicostata (Ritt.) Backeb.

In Chile cacti are distributed from approximately the 35° latitude south towards the equator. North of 30° latitude south (from the town of La Serena) they are the only higher plants to be found, along with several species of xerophilous plants. However, their ranks are not as diverse as in Argentina or Mexico. Apart from a number of opuntias and cerei, there are basically two characteristic groups — cacti of the genus *Copiapoa* and cacti classed in the genera *Neochilenia, Neoporteria* and *Horridocactus*, which could be combined into a single genus. Even though most of these cacti are very interesting in themselves, when massed in a comprehensive collection the effect is rather bland. One species that definitely breaks the monotony is the pale blue-grey *Neochilenia paucicostata*, found, together with several closely related species, north of the town of Antofagasta. Young specimens have a globose body which becomes more columnar in maturity, reaching a height of up to 20 cm (8 in), at which time it measures about 10 cm (4 in) in diameter. The spines are black when they emerge, later turning grey; the central spines are about 4 cm ($1\frac{1}{2}$ in) long. *N. paucicostata* flowers readily and repeatedly as a small seedling even in collections. The flowers, coloured pinkish-white and 5 cm (2 in) across at the most, are not very striking. Like other members of the genus, *N. paucicostata* is not difficult to grow. Nevertheless it is often grafted, for example on *Cereus peruvianus*, on which it produces prolific spines and soon attains flower-bearing size. Later, however, it loses its natural appearance because it grows too tall and therefore growing it on its own roots may be preferred and can be recommended. In summer it should be put in a warm, sunny spot and in winter it should be provided with cool conditions (a temperature of less than 10°C/550°F) because all Chilean cacti tend to start growth ˙maturely in higher winter temperatures.

3

Neochilenia paucicostata (1) is coloured grey-blue. The variety *viridis* (2), as its name indicates, has a green skin. This variety is very similar to *N. hankeana*, until the 1950s the only known and validly described neochilenia from this region. At a later date F. Ritter found several further species here, all more or less related to *N. hankeana.* The most interesting is *N. floccosa* (3), which has white wool and hairs on the crown of flower-bearing specimens.

2

1

165

This attractive Chilean species belongs to the group of so-called inland neoporterias characterized by dense colourful spines. *N. multicolor* has a globose stem at first, which later becomes shortly cylindrical, reaching a maximum height of 20 cm (8 in). The colour of the spines of individual specimens ranges from pale to dark. The spines also differ markedly in strength as well as shape. In some forms they are very soft and numerous, in others they are stronger. *N. multicolor* bears flowers from the latest areoles on the crown in spring and often again in late summer. The flowers are 6—8 cm ($2\frac{1}{3}$—3 in) long and about 5 cm (2 in) across, with long-pointed perianth segments coloured purplish-violet. Each flower blooms for almost a whole week, remaining open even at night and in rainy weather. This species, like other members of the genus, can be grown on its own roots, but, naturally, must be provided with conditions similar to those of its native habitat, i.e. a free-draining mineral substrate and in summer as much direct sunlight as possible and only slight, occasional watering. Although its growth is slower, it does not grow too tall and produces a dense covering of spines. Novice cactus growers are advised to graft this species on *Cereus peruvianus* or *Eriocereus jusbertii* stock. In winter *N. multicolor* requires an absolutely dry substrate and a cool temperature (around 10°C/550°F), otherwise it begins growth prematurely, thereby distorting its shape.

2

Neoporteria multicolor (1) takes its specific name from its spines, which differ markedly in colour among various specimens. Besides the commonest colour — straw yellow — the spines may be nearly white or brown (2) to black. *N. gerocephala* (3) is a related species also native to central Chile where its range links up with that of *N. multicolor.* It is sometimes designated by the invalid name of *Neoporteria sinilis.*

Its spines are generally grey or greyish with a darker brown to nearly black central spine. Occasionally encountered are specimens with pale or sometimes pure white spines.

3

1

Notocactus apricus (Arech.) Berger

The genus *Notocactus* numbers some 150 species along with numerous varieties and forms. The species are classed in the following subgenera: *Brazilicactus, Eriocactus, Notobrasilia, Notocactus, Neonotocactus* and *Malacocarpus*. Some of these subgenera are often considered to be separate, independent genera. All species grow in the flat to hilly country of Uruguay, Paraguay and the neighbouring territories of southern Brazil and Argentina. They have rather small globose to cylindrical stems and, generally, yellow flowers with a red stigma.

N. apricus was described in 1905 by the Uruguayan botanist J. Arechavaleta, who named it *Echinocactus apricus*. It is a globose cactus up to 10 cm (4 in) across with a bright green stem covered by a tangle of yellow spines, up to 1.5 cm ($\frac{1}{2}$ in) long, that later turn greyish. The flowers appear close to the vegetative centre and open fully only at noon over a period of about three days. They are relatively large in proportion to the rather small body, measuring 8 cm (3 in) in length and the same across. Whereas the perianth segments are yellow, the stigma is a glowing red. Like most notocacti, *N. apricus* is easy to grow. Nevertheless, it is necessary to bear in mind the soil and climatic conditions of its native habitat and try to approximate them. The substrate should contain river sand, rather light, slightly acidic, nourishing soil and sufficient humus. During the growing period the plant should be watered more frequently because the root system regenerates with great difficulty if it remains dry for a long time. Even though the word *apricus* means sun-loving, *N. apricus* requires a slightly shaded situation in summer and in winter a dry substrate and a temperature of 5—15°C (41—59°F).

2

Specimens of *Notocactus apricus* (1) in collections generally correspond to the original description with their yellow spines. Besides yellow-spined individuals, however, there are also some with dark brown spines (2), designated var. *nigrispinus*. *N. concinnus* (3) is a closely related species and sometimes *N. apricus* is classed as a mere variety of this. The two can be distinguished primarily by the covering of spines on the crown. In *N. apricus* the

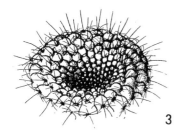

crown is covered with intertwined spines; in *N. concinnus* the area around the vegetative centre at the top is bare and the overall covering of spines is less dense. The large yellow flowers, up to 10 cm (4 in) across, resemble those of *N. apricus.*

3

1

Notocactus graessneri (K. Schum.) Berger

This golden-spined cactus is often classed in the genus *Brasilicactus*, originally established for two species differing rather markedly from the other notocacti. According to the latest concept, it is classed in the subgenus *Brasilicactus*, which was established within the framework of the genus *Notocactus*. *N. graessneri* is found in a number of localities in the state of Rio Grande do Sul in southern Brazil, in other words in a region with abundant rainfall. It grows in the shallow, free-draining, rapidly drying soils of rock outcrops, that prevent lengthy waterlogging of the plant's root system as well as the growth of competitive vegetation. In mature age it attains a diameter of approximately 10 cm (4 in) and often produces offshoots, forming clusters up to 50 cm ($19\frac{1}{2}$ in) across. Whereas in some localities the plants have rather globe-shaped stems, in other plants the stems are slightly elongated. The crown, however, is always at a slant. The stem is covered by a dense tangle of deep yellow spines about 2 cm ($\frac{3}{4}$ in) long. There are about 60 in each areole. Also noteworthy are the flowers that form a wreath around the vegetative centre at the top. They are glowing green to yellow-green and approximately 2 cm ($\frac{3}{4}$ in) long. They are still green when they emerge, later acquiring a yellow tinge, and remain open for many days regardless of the time of day or the weather. The spiny, globose fruits are small, as are the seeds they contain. *N. graessneri* is not difficult to grow; it does well on its own roots. The only thing it does not tolerate is a heavy, poorly draining substrate. During the growing period it requires a sunny, partly shaded situation. In winter, small seedlings should be watered at least once or twice so that they do not dry up completely.

2

Notocactus graessneri (1) has several varieties. Readily distinguished is var. *albisetus* (2), named after its white, bristly hairs. Whereas *N. graessneri* has yellow spines, those of var. *albisetus* are yellowish-white to nearly white and each of its areoles bears a greater number of white, bristly hairs up to 3 cm (1 in) long. Var. *albisetus*, with its pale spines, resembles the other member of the subgenus *Brasilicactus* — the species

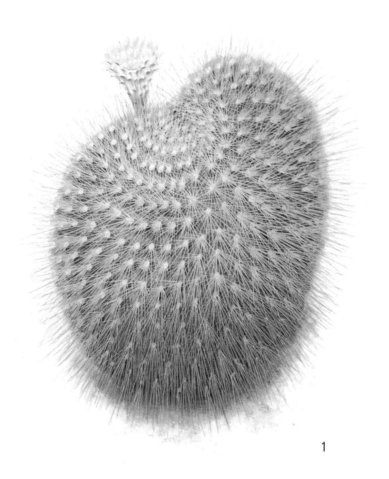

1

3

N. haselbergii (3). They may be mistaken for one another only until they flower, however. The typically fire-red flowers of *N. haselbergii* are generally produced from early spring until autumn. They are 1.5 cm ($\frac{1}{2}$ in) long and 1 cm ($\frac{1}{3}$ in) across. Their great advantage is their long period of flowering — almost a whole month — during which time they remain open even at night and in rainy weather.

Notocactus horstii Ritt.

Native to the state of Rio Grande do Sul in southern Brazil, or rather more precisely to the southern slopes of the Sierra Geral and the region south of this mountain range, are several notocacti that are very similar to one another but may be readily distinguished. It was the atypical colour of the flowers, whose range includes various shades of orange-yellow, red and violet, that attracted the attention of cactus growers. The first of this group to be discovered and described was *N. horstii.* In mature age it reaches a good size, often up to 30 cm (12 in) in breadth and 1 m (3 ft) in height. It generally remains solitary, only occasionally producing offshoots. The radial, pale brown to almost white spines are 1—3 cm ($\frac{1}{2}$—1 in) long, the central spines slightly longer and mostly brown. In collections, *N. horstii* flowers in late summer; only very occasionally does it bear flowers sooner. These are 3—3.5 cm (1—$1\frac{1}{2}$ in) long and 3 cm (1 in) across; the perianth segments are orange-yellow with orange-red to vermilion tips. The fruits generally ripen in the spring of the following year. *N. horstii* is moderately difficult to grow. It is mainly the small seedllings that pose the greatest problem for they have a tendency to dry up in winter and should therefore be watered moderately about twice during this period. In summer, the plants should be provided with a warm, partly shaded situation, moist atmosphere and an adequately humus-rich and nourishing substrate.

3

Notocactus horstii (1) is variable not only in the colour of the spines but also in the colour of the perianth segments. The flowers may be orange, which is the commonest colour, as well as nearly yellow or vermilion. A population with purple to violet flowers (2) was discovered in one totally isolated locality. These plants were described in 1979 by C. Ritter and given the apt name *N. horstii* var. *purpureiflorus.* Closely related is the species *N. purpureus* (3). As the name indicates, it, too, is a purple-flowering notocactus. In general appearance, however, it is quite

1

2

different from both *N. horstii* and its
variety *purpureiflorus.* It grows more in
width at first, its skin is a bright pale
green, and it has larger areoles with
a dense covering of wool that, in adult
specimens, extends along the edges of
the ribs in the form of a continuous
felted border.

Notocactus leninghausii (Haage jr.) Berger

N. leninghausii, named in honour of the Brazilian cactus grower F. G. Lenninghaus, is native to the state of Rio Grande do Sul in Brazil. Botanists differ in their opinions as to the classification of this species and all other eriocacti. Whereas some systematists class these columnar, finely spined cacti with oblique crowns and typical reproductive organs in the separate, independent genus *Eriocactus*, others class them in the subgenus *Eriocactus* within the genus *Notocactus*. *N. leninghausii* has a globe-shaped stem at first, that later becomes cylindrical to columnar and produces offshoots at the base. In older age it attains a height of more than 1 m (3 ft) and a diameter of 10 cm (4 in). The areoles bear up to fifteen whitish-yellow radial spines and three to four central spines coloured golden-yellow. The autogamous flowers, about 5 cm (2 in) across, are the same colour. Also encountered in collections is the mutant *N. leninghausii* f. *apellii*, which has a globose stem producing prolific offshoots even in youth, and short spines. Its opposite is var. *longispinus*, which has central spines up to 5.5 cm (2 in) long and a smaller body, for which reason it is sometimes designated as var. *minor*. *N. leninghausii* is easy to grow, tolerating minor mistakes on the part of the grower. It forms thick columns, densely covered with spines only in a sunny and warm location and in a nourishing, humus-rich substrate. In winter the temperature should not drop permanently below 5—10°C (41—50°F). On reaching maturity, when the crown begins to slant, it should be placed at the same angle if moved to another spot, so that the crown does not turn towards the sun, thereby deforming the top of the stem.

2

Whereas the flowers of most cacti close for the night or in rainy weather, the flowers of *Notocactus leninghausii* (1) remain open for four to five days, regardless of the time of day or the weather. Typical of all members of the subgenus *Eriocactus* is the slanting angle of the crown (2). However, the crown starts to slant only after the plant has reached flower-bearing size. It slants in a south-facing direction and the sun's

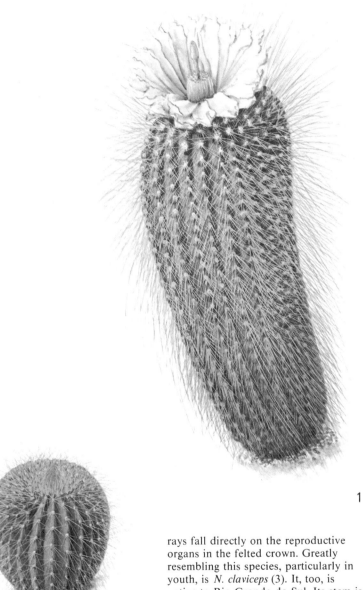

1

3

rays fall directly on the reproductive organs in the felted crown. Greatly resembling this species, particularly in youth, is *N. claviceps* (3). It, too, is native to Rio Grande do Sul. Its stem is broadly club-shaped, up to 50 cm (19 $\frac{1}{2}$ in) high and 20 cm (8 in) across, has fewer spines and produces offshoots. The flowers are 5.5 cm (2 in) long and coloured pale sulphur-yellow.

175

N. *magnificus* occurs in relative abundance on a not very large area in the Sierra Geral in the south Brazilian state of Rio Grande do Sul, where it grows in free-draining, stony, moderately acidic soils in a region with relatively large amounts of rainfall. In mature age it reaches about 50 cm ($19\frac{1}{2}$ in) in height and 25 cm (10 in) in breadth, and often forms huge clumps. The skin is green at first; in adult specimens it is blue-green. There are generally eleven to fifteen ribs, 1.5—3 cm ($\frac{1}{2}$—1 in) high. The extremely woolly areoles of adult plants nearly touch one another and form a felted border along the edge of the ribs. Whereas the central spines are softly bristly and coloured pale yellow to brown, the radial spines are like fine hairs and white in colour. The pale yellow flowers emerge on the slanting crowns of plants when they attain a diameter of 7 cm ($2\frac{3}{4}$ in). They measure about 5 cm (2 in) across and remain open even at night for about five days. N. *magnificus*, like other members of the subgenus *Eriocactus*, is rather difficult to grow in the early stage when the minute seeds develop into small seedlings that must be watered now and then during the winter. In ensuing years their growth is rapid, so that grafting is quite unnecessary. N. *magnificus* requires a warm and sunny situation that is slightly shaded at least at the beginning of the growing period and when growth stops. In winter it does not stand up well to drops in temperature below 10°C (50°C), when, especially in the case of greater atmospheric moisture, black spots caused by a fungus will appear on the skin. Propagation is by means of seeds as well as offshoots, which root readily.

2

Notocactus magnificus (1) differs from the other species of the subgenus *Eriocactus* by having broad, prominent ribs and a blue-green skin. Its reproductive organs, i.e. flowers, fruits and seeds, are identical to hitherto known eriocacti. The fruit (2) is globe-shaped and opens in a manner quite different from the fruits of other notocacti. At maturity, it splits in a circle at the base round the point of attachment to the areole, so that it is

almost entirely separated and the seeds drop out of the bottom. Found in the same region is the closely related *N. warasii* (3), which reaches a breadth of 13—15 cm (5—6 in) and a height of 50 cm ($19\frac{1}{2}$ in), has a greater number of ribs and dark green skin, and begins bearing flowers relatively early, usually at the age of six years. The flowers are pale yellow and practically indistinguishable from those of *N. magnificus.*

3

1

Notocactus ottonis (LEHM.) BERGER

N. ottonis has been kept in European collections for the past 150 years. It is an undemanding plant that is easily propagated by vegetative means and bears large yellow flowers readily and early on. It is often cultivated in collections because the extensive localities of its native habitat are readily accessible. It occurs in a number of forms and varieties, differing in a lesser or greater degree, in territories of four South American countries: Brazil, Uruguay, Paraguay, and Argentina. The body is globose, 5—11 cm (2—$4\frac{1}{3}$ in) across and slightly elongated in adult specimens. The ribs are broad and rounded. The prominent, white-felted areoles bear ten to eighteen radiating needle-like radial spines and three to four central spines. The central spines are slightly thicker and longer and coloured brown to brownish-red. The autogamous flowers are yellow with a silky sheen, measure about 6 cm ($2\frac{1}{3}$ in) across, and open in the afternoon over a period of three to four days. An interesting phenomenon is the irritability of the stamens, which is well developed in this species and its varieties. If touched, the stamens bend towards the stigma within one second and pollinate it. In the wild, the stamens are irritated either by insects or by the wind; in collections this reaction may be provoked by touching them with forceps. *N. ottonis*, like all closely related species, is readily propagated by offshoots as well as seeds. Offshoots are formed around every older plant, growing underground at first. During this period they generally form rootlets so they need not be put to root when they are separated from the parent plant. Because *N. ottonis* does not tolerate alkaline soil, in which its growth is slower and it loses its roots and becomes corky at the base, it should be grown in a light, humus-rich substrate with a greater proportion of peat and leaf mould.

2

Notocactus ottonis (1) flowers profusely and repeatedly. While spent flowers are developing into fruits, the plant forms further buds and new flowers open. This cactus has several dozen forms, varieties and closely related species. Most attractive is var. *vencluianus* (2), with glowing red to orange-red flowers. The anthers are generally without pollen so it reproduces mostly by vegetative means. Of the many closely related species, *N. acutus* (3), with a dark green skin, sharp ribs and dark brown to black, twisted spines, is noteworthy.

3

1

N. rutilans is one of the smallest members of this genus and differs from the rest in having violet-tinted flowers. It is native to Uruguay, where it grows on partly grassy hills near the Brazilian border. It was discovered here in 1936 but was not described until 1948, for it long remained unclear whether it was a separate species or a violet-flowering variety of *N. mueller-melchersii*, which also grows in this region. *N. rutilans* has a globe-shaped body at first, which later becomes slightly elongated, is 5—10 cm (2—4 in) long and does not produce offshoots. Each areole bears fourteen to sixteen radial spines but only two central spines, coloured brownish-red and measuring 0.8 mm to 3 cm (1 in) in length; the lower of the two central spines, directed downwards towards the base of the stem, is thicker and longer than the other. Specimens with long central spines are designated var. *longispinus*, i.e. long-spined. The pale violet flowers, arising on the crown of the seedling as early as the fifth year, are up to 4 cm ($1\frac{1}{2}$ in) long and 6 cm ($2\frac{1}{3}$ in) across. *N. rutilans* grows well on its own roots. It requires a light, sandy-humus substrate. Grafting is the recommended choice only in the case of the dwarf form *storianus*. During the winter, small seedlings should be watered occasionally, but not older plants; although these dry up quite a bit during the winter, they soon regain their original shape after the first application of water in early spring. During the growing period a moderately shaded situation best suits their needs, and in winter a temperature of 8—15°C (46—59°F).

2

Notocactus rutilans (1) has quite unusual flowers compared with those of other notocacti. The perianth segments are a pinkish-carmine that becomes paler towards the centre, which is yellowish. The buds emerge on the crown in early spring and, as in most cacti, the development of the flowers takes about five weeks, in cooler weather even longer. They remain closed at night and in the forenoon (2), opening only in the

afternoon over a period of two to three days. A mutant of this species, described as *N. rutilans* f. *storianus* (3), was later propagated in large numbers by vegetative means. Characteristic features are its dwarf habit, short spines and prolific production of offshoots. In order to obtain profuse and regular flowering, it is recommended that one removes new offshoots as they emerge.

1

3

Notocactus scopa (Spreng.) Berger

N. scopa is one of the loveliest members of this genus. Its appearance is attractive from earliest youth, and in mature age the profuse flowering adds to its attraction. It is distributed in a number of localities in the moderately hilly territory of Uruguay and southern Brazil. It generally remains solitary, only very occasionally producing offshoots. The stem is globose at first, later slightly elongated. In mature age it reaches a height of about 30 cm (12 in) and a breadth of 10 cm (4 in). Each areole bears a large number (about 40) of needle-like, glassy-white radial spines 5—7 mm (2 — 2 $\frac{1}{2}$ in) long, but only three to four central spines, which are slightly thicker and longer and generally coloured brownish-red. The flowers arise on the crown of older specimens, when the stem begins to lengthen slightly. Often several emerge simultaneously, not just once but a number of times during the growing period. They are about 5 cm (2 in) across, pale yellow with a silky sheen, and open in the afternoon over a period of four days. *N. scopa* is an undemanding species that grows well even on its own roots. Like other cacti with slender, much-branching roots, it requires a light, sandy-humus substrate. Because it is native to a region with relatively frequent rainfall, it should be watered regularly during the growing period. Small seedlings in particular do not stand up well to lengthier dry spells and should be watered occasionally even in winter. *N. scopa* flowers profusely and repeatedly and produces a thick covering of spines if placed in a warm, very sunny spot.

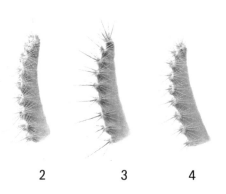

2 3 4

The central spines of *Notocactus scopa* (1) are generally coloured brownish-red but there are numerous forms with central spines of a slightly different or pronouncedly different colour. The spines, arranged in a thick bundle in each areole, resemble a broom, whence the species takes its name (*scopa* means broom). Specimens with nearly white spines are designated f. *candidus* (2), with brown spines f. *bruneispinus* (3), with yellow spines f. *daenikerianus* (4). *N. sucineus* (5) is a closely related species. Its spines are golden-yellow, like amber, hence the name *sucineus*, which means amber-like. Some specimens have pale yellow to white spines and these are called f. *albispinus*. *N. sucineus* grows in southern Brazil in the state of Rio Grande do Sul, and conditions for its cultivation are the same as for *N. scopa*.

5

1

Notocactus uebelmannianus Buin.

The state of Rio Grande do Sul in southern Brazil long escaped the attention of cactus collectors. Not until the second half of this century, more specifically the 1970s, did a detailed investigation of this area bring surprising finds. One of the new, entirely different and interesting discoveries was *N. uebelmannianus.* When adult, it measures up to 15 cm (6 in) across, thereby ranking among the largest of the notocacti. Its stem is flat-topped and globose, with ribs divided into flat tubercles. The radial spines, up to 3 cm (1 in) long, are pressed close to the body, the single central spine, if present (it may be absent), is directed downwards towards the base of the stem. The felted areoles, with six spines spread out sidewards from the centre, resemble spiders. The short funnel-shaped flowers, up to 5 cm (2 in) across, are a gleaming wine-red colour. It is interesting to note that in the type locality about 85 per cent of the plants had flowers of this colour; the remaining 15 per cent had yellow flowers. Yellow-flowered specimens were described as the form *flaviflorus.* In its native habitat *N. uebelmannianus* grows on hills amid rocks and only some specimens are partly shaded by shrubs or by large bromeliads. In cultivation they therefore require a sunny, moderately shaded situation and liberal watering during the growing period. In winter, a light application of water should be supplied occasionally only to one- and two-year-old seedlings. This cactus does not produce offshoots and is propagated only by means of seeds.

2

The flowers of *Notocactus uebelmannianus* (1) range in colour from pale to deep violet. Their colour, as well as size, changes during the course of flowering. When they open, the flowers are always a darker hue and approximately one-third smaller than in the ensuing days. Increase in size and fading of colour during flowering are not exclusive characteristics of this species. They may be encountered in

1

other cacti as well, but are not always so
pronounced. Related species native to
Rio Grade do Sul include
N. crassigibbus (2), which resembles
N. uebelmannianus in habit but has
yellow flowers up to 10 cm (4 in) across
and ribs divided into prominent
tubercles, and *N. arachnites* (3), with
yellow flowers only about 4 cm ($1\frac{1}{2}$ in)
across and dense spines — those on the
crown form a tangled mass.

3

Oreocereus trollii (KUPP.) BACKEB.

The genus *Oreocereus* comprises several species, all very attractive, of columnar habit and with a thick covering of hair-like spines.

O. trollii is the smallest and also the loveliest species. It is found in the Andes of Bolivia and northern Argentina at elevations of 3,500—4,000 m (11,480—13,000 ft), where it grows on stony substrates without cover of taller vegetation. At maturity it produces off-shoots and forms small clusters. The individual stems grow up to 50 cm (19½ in) high and 10 cm (4 in) across. The white, hair-like spines, up to 7 cm (2¾ in) long, are greatly intertwined from early youth. Besides hair-like radial spines, each areole bears one or sometimes more large central spines coloured yellow or reddish-brown. The flowers, arising near the crown of older plants, are narrow, pinkish-red to carmine, and remain open even at night. *O. trollii* produces offshoots only in mature age and is thus propagated almost exclusively by means of seeds. Small seedlings, as well as adult plants, should be overwintered in cool conditions and in a dry substrate, otherwise their crowns become deformed. When grafted on *Cereus peruvianus, O. trollii* produces dense white hairs and large central spines at a young age. To obtain broad stems thickly covered with spines, this species should be grown in a sunny spot that is frequently aired during the period of peak summer and autumn night temperatures. In winter, it requires the least possible atmospheric moisture and a temperature of about 10°C (50°F).

2

Oreocereus trollii (1) has very attractive white, hair-like spines that shade the skin, protecting it against strong sunlight as well as reducing the evaporation of water during the resting period. Another high mountain oreocereus that is native to Bolivia and Argentina is *O. celsianus* (2). This is a larger cactus with a stem up to 15 cm (6 in) across and 2—3 m (6½ — 10 ft) high and a sparser covering of white, hair-like spines. It shows

1

a marked variability not only in the
length of its stem, but also in the colour
of the central spines, which may range
from amber-yellow to brownish-red.
Cultivation is the same as for *O. trollii*.

Oroya peruviana (K. Schum.) Britt. et Rose

O. peruviana, like other members of the genus *Oroya*, is a typical high-mountain plant. It is found in the Andes of Peru at elevations of about 4,000 m (13,000 ft), which makes it one of the highest-growing cacti. It is prized by collectors primarily for the unusual colour of its flowers. The stem is globose, 10—20 cm (4—8 in) across, with a glossy green skin. It sometimes produces offshoots. The ribs, usually 12—21 in number, are divided, in lesser or greater degree, by transverse grooves into rounded tubercles. The flowers form a wreath near the crown. The buds emerge at the end of winter and are among the first cacti buds to open in early spring. The flowers are 2—3 cm ($\frac{3}{4}$—1 in) long and coloured a glowing purplish-red with a conspicuous yellow throat. In the neighbourhood of the mining town of Oroya, from which the genus takes its name, and further south, *O. peruviana* occurs in numerous local forms, many of which are classed as varieties. Examples are var. *neoperuviana*, which is larger and has denser spines, var. *depressa*, which has a flat stem and prominent tubercles, and var. *conaicensis* and var. *citriflora*, both of which have deep yellow flowers. *O. peruviana* is best grown without glass in summer, for example in an open frame, and in the greenhouse it should be placed close to the windows that are used to air the premises, for a sultry atmosphere is the most damaging of all; it then grows too tall, has sparser spines, and the skin is disfigured by unattractive spots. It is one of the cacti that is most susceptible to sun scorch and must therefore be provided with shade and ample ventilation during the entire growing period. All species of this genus become corky at the base in age and this should be viewed as a natural development.

3

The spines of *Oroya peruviana* (1) vary in density and length as well as coloration. Commonest are plants with yellow spines, but one may also come across specimens with white, red or brown to black spines (2). The genus *Oroya* comprises only a few species. Whereas *O. peruviana* is most easily grown and hence the species most often

1

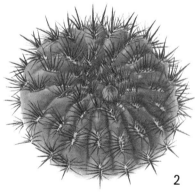

2

grown, *O. borchersii* (3) is considered the most attractive of them all, with its amber-yellow spines and vivid green skin. Also very attractive is the variety *fuscata*, with reddish-brown spines, and the form *aureotenuispina*, with pale green skin and long, slender, more prominently jutting spines. *O. borchersii* does not flower readily in collections.

189

Parodia maassii (HESSE) BERGER

The genus *Parodia* comprises readily and profusely flowering species with colourful spines. They are mountain plants found growing in the Cordilleras of northern Argentina and Brazil. Some 100 species have been described to date.

P. maassii belongs to the group of so-called long-spined parodias. It is native to southern Brazil and northern Argentina, where it grows up to elevations as high as 3,600 m (11,800 ft) above sea level. It is one of the highest-growing, and at the same time one of the commonest, of parodias. The body is spherical to slightly elongated and up to 15 cm (6 in) across in adult forms. The ribs are slightly tuberculate and spirally arranged. The eight to ten radial spines, 5—10 mm long, are honey-yellow at first, later whitish. The four central spines are stronger, irregularly curved to hooked, and arranged in the form of a cross; the lowermost spine is the longest, sometimes more than 3 cm (1 in) long, and pale brown. The flowers are 3.5 cm (1½ in) across and coloured salmon-red. *P. maassii* has several varieties. Readily distinguished, for example, is var. *albescens* with pale central spines that are nearly straight in mature age. *P. maassii* has large seeds so there are no particular problems with sowing. Seedlings, as well as adult plants, do not tolerate a humid and stuffy atmosphere and must therefore be provided with ample ventilation. Like all other long-spined parodias, this species is rather susceptible to sun scorch. In its native habitat it grows in the shade of pine trees and thus the most suitable situation for this cactus is one that is warm and adequately shaded. Even though it is a mountain species, it does not stand up well to low winter temperatures. In the case of exposure to a temperature of less than 5°C (41°F) for a lengthy period, accompanied by higher atmospheric moisture, unattractive spots, caused by a fungus, appear on the skin.

2

3

Parodia maassii (1) is distributed over a wide range and that is one of the reasons for the variability of this species. Although the original description states that the length of the central spines is 3 cm (1 in), specimens with 5- to 8-cm-long (2—3 in), variously curved spines were later discovered. The central spines may be greatly curved, nearly straight (2), or hooked (3). The closely related *P. suprema* (4) has brown to black central spines that turn grey in age, and scarlet flowers. There is a great resemblance between it and the dark-spined *P. maassii* var. *intermedia*, for which it might readily be mistaken. However, *P. maassii* var. *intermedia* grows at lower elevations and has dark red flowers and slightly longer central spines (4—6 cm/$1\frac{1}{2}$ — $2\frac{1}{3}$ in).

4

1

Parodia mutabilis BACKEB.

Parodia mutabilis is currently one of the commonest parodias in collections. In great measure, it owes its popularity to its large, bright yellow flowers, which arise repeatedly near the crown at the beginning and end of summer. Adult plants, in particular, produce whole series of flowers, opening simultaneously over a period of about one week. This species is native to the mountain regions of northern Argentina where, in the province of Salta, it grows on stony, rather acidic soils unshaded by taller vegetation. *P. mutabilis* does not produce offshoots. Its stem is about 8 cm (3 in) across, spherical at first, later becoming more columnar in mature age. The radial spines, about 50 to each areole, are thin, almost hair-like, and white. One of the four central spines is hooked, about 1 cm ($\frac{1}{3}$ in) long and coloured red to orange-brown. Its colour is very variable, which is reflected in the name of the species (*mutabilis* means changeable). C. Backeberg described several varieties based on the colour of the central spine, but there is no justification for this from the botanical viewpoint for they are not from isolated localities. He designated specimens with reddish-brown spines as var. *carneospina*, those with rusty-red spines as var. *ferruginea*, and those with finer, thinner central spines as var. *elegans.*

In the growing period, *P. mutabilis* requires an unshaded, airy situation and more liberal watering, particularly in spring and late summer. The plant should be placed further in from the edge of the ledge or shelf, for otherwise the hooked spines might catch in one's clothing and the plant will then be readily lifted from the substrate in which it is only lightly rooted.

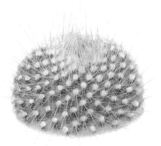

2

Parodia mutabilis (1) has pale to golden-yellow flowers about 3.5 cm (1$\frac{1}{3}$ in) across. They open in the afternoon and close at sundown. Also native to the province of Salta is the species *P. aureispina* (2), likewise with a great number of fine radial spines and several stronger central spines. One central spine is directed downwards and

1

hooked to a greater or lesser degree. The
plant's specific name is derived from the
colour of the central spines (*aureispinus*
means golden-spined). At first glance the
yellow flowers are indistinguishable
from those of *P. mutabilis.* The species is
easy to grow and flowers profusely even
in the collections of novice cactus
growers.

Parodia nivosa Frič et Backeb.

P. nivosa is one of the loveliest discoveries of the Czech collector of cacti A. V. Frič, found by him in 1928 on his last field trip in the province of Salta, northern Argentina, where it grows at elevations of 2,000—2,500 m (6,500—8,200 ft). The attractive combination of snow-white spines and bright red flowers makes it not only the loveliest of parodias but the most beautiful of all cacti. It does not cluster but remains solitary. The globose body begins to lengthen in age and may be as much as 15 cm (6 in) high when it measures 8 cm (3 in) across. The skin is pale olive-green. The ribs are composed of spirally arranged conical tubercles. The areoles on the crown are white-felted; in older areoles the wool thins somewhat. The radial spines, eighteen or more in each areole, are glassy white, nearly transparent and 6—10 mm long, arranged in a radiating pattern, thinly needle-like, fragile and brittle. The four central spines, arranged in the form of a cross, are stronger than the radial spines and are straight, 2—2.5 cm (¾—1 in) long and the same colour as the radial spines. Only occasionally are the spines on the crown brownish at the base, however this colouring soon disappears. The flowers, 4—5 cm (1½—2 in) across, arise on the crown, generally several at a time, and are various shades of bright red. *P. nivosa* is easy to grow. The only problem is in the propagation of the small seeds. The seedlings that sprout from these seeds are minute and more pronounced growth does not set on until the second and third year after sowing. In view of the plant's less extensive root system, it should be put in a small pot. In winter, *P. nivosa* requires an absolutely dry substrate and a temperature of about 5—10°C (41—50°F).

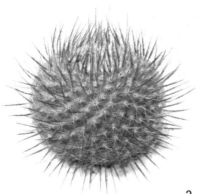

Parodia nivosa (1) takes its name from the white spines that completely cover the stem (*nivosus* means snow-white). *P. faustiana* (2) is a closely related species. It, too, is native to the province of Salta and could thus be considered rather as a variety of *P. nivosa* than a separate independent species. The

2

1

areoles bear about twenty radial spines
that are glassy white and 1 cm ($\frac{1}{3}$ in)
long, and four stronger central spines,
about 2.5 cm (1 in) long, coloured
brown to black. The flowers resemble
those of *P. nivosa* and repeatedly arise
close to the crown during the spring and
summer months.

Parodia penicillata FECHSER ET STEEG.

This parodia was discovered by H. Fechser, the well-known collector of cacti, in 1951 in the southern part of the north Argentine province of Salta in the mountains near the town of Cafayate. At first glance, it resembled the known species *P. chrysacanthion*, from which it differed in its emerging spines arranged in a brush-like bundle in the areoles. *P. penicillata* grows solitarily. The stem, globose at first, later columnar, is 12 cm (4¾ in) across at the most and 30 cm (12 in) high (sometimes more), and lies on the ground or hangs down from rocks. Both the central and radial spines are yellow to whitish-yellow, but in some specimens glassy white or dark yellow to brown. They are very numerous and up to 5 cm (2 in) long. The bright red flowers are about 4 cm (1½ in) across. C. Backeberg described two varieties on the basis of their differently coloured spines. Although there is cause for doubt as to their validation botanically, there is no denying their importance from the collector's viewpoint. White-spined plants are designated as var. *nivosa* and plants with reddish-brown spines as var. *fulviceps*. *P. penicillata* is not particularly hard to grow. The only problem is with the sowing of the small seeds and the resulting minute seedlings which grow very slowly during the first two years. It is recommended that adult plants be watered at greater intervals, for example once in three weeks. This will decrease the risk of them dying and will result in denser, more colourful and longer spines than if they are watered frequently.

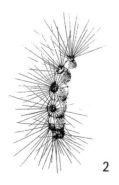

Parodia penicillata (1) flowers readily and generally produces whole series at a time as well as several times during the growing period. The flowers are orange-red. *P. penicillata* derives its name from the arrangement of spines in the areole (2) resembling a brush (*penicillata* means brush-like).
P. chrysacanthion (3) has golden-yellow spines that are straight and numerous, like those of *P. penicillata*. The two species can be readily distinguished during the flowering period. The pale yellow, bell-like flowers of

P. chrysacanthion, measuring 2.5 cm
(1 in) across, arise from the woolly
crown at winter's end or in early spring.
They open even in inclement weather,
such as rain. The pollen is generally
insufficiently ripe because of the low
temperatures and so pollination of the
flowers is often unsuccessful.

3

1

Rebutia heliosa RAUSCH

The genus *Rebutia* comprises some 100 species found in the eastern Cordilleras at elevations of 1,500—4,000 m (5,000—13,000 ft). They are miniature, soft-fleshed, generally clustering plants that flower readily and profusely at an early age.

R. *heliosa* is native to the mountain regions of Bolivia, to the province of Tarija, where it grows at elevations of 2,400—2,500 m (7,880—8,200 ft) on stony substrates unshaded by taller vegetation. At first glance it resembles certain miniature cacti of Mexico, but it has far lovelier flowers and is much easier to grow. The body of R. *heliosa* is flat-topped and globose to columnar, 1.5—2 cm ($\frac{1}{2}$—$\frac{3}{4}$ in) across and 2 cm ($\frac{3}{4}$ in) high. The numerous white spines are pectinate and only 1 mm long. There are generally 24—26 on each areole. The areoles are covered with dark brown felt. The flowers emerge at the base of the stem, and are 4.5—5.5 cm ($1\frac{3}{4}$—2 in) long and 4 cm ($1\frac{1}{2}$ in) across and coloured orange to fiery-red. Although R. *heliosa* is often grafted, it grows relatively well even on its own roots. It should be provided with a mineral, relatively free-draining substrate and watered liberally when the substrate is completely dry. Its fullest beauty will be attained if it is placed in a sunny and airy spot. Although it is also propagated by means of offshoots, propagation from seed is better, for the seedlings form a strong turnip-like root early on, which is hardier than the finely branched rootlets of a shoot that has been cut off the parent plant. Because R. *heliosa* immediately starts growth at a higher temperature when the substrate is moist, it should be overwintered in an absolutely dry substrate and in a cool room at a temperature of 5—10°C (41—50°F).

Even though *Rebutia heliosa* (1) remains solitary during the first few years, at a later age it forms large clusters of numerous small stems. The short, dense spines (2) provide the soft bodies with good protection against sun scorch. R. *heliosa* has several related species. Their mutual relationship is attested to primarily by the structure of the flowers, which have a long slender tube. Whereas

4

some species have only slightly different spines, for example *R. albopectinata* (3), others differ in their overall habit and the colour of the flowers, such as the solitary *R. narvaezensis* (4), which differs from most of the other species in its pale pink to white flowers, which it begins to produce at the age of two years, when the stem measures 1.5 cm ($\frac{1}{2}$ in) in diameter.

2

3

1

Rebutia krainziana KESSELR.

Species of the genus *Rebutia* are rewarding plants from the cactus grower's viewpoint in that they flower readily at an early age when they measure only a few centimetres. Most flower quite early in spring so that larger collections of these plants are a sight to behold at that time of the year, although in the ensuing summer months they present a rather bland picture without their blossoms. One of the few exceptions is *R. krainziana*, which brightens such a uniform collection of rebutias in summer not only with its unusual habit, but also with its deep red flowers, which open at this time. Although native to Bolivia, according to W. Kesselring's description, it has not been found again in the wild since he discovered it. *R. krainziana* has a globose stem that produces offshoots, but not until it measures 4 cm ($1\frac{1}{2}$ in) in diameter, which is generally the most its soft-fleshed bodies can attain. The skin is a dull green, tinged violet in full sunlight. The striking, white-felted areoles bear eight to twelve delicate, thin, straight spines about 2 mm long. The allogamous flowers are among the loveliest in the whole genus. They are large in proportion to the plant's small size — about 4 cm ($1\frac{1}{2}$ in) across and 3 cm (1 in) long — and coloured a vivid red. *R. krainziana* is readily propagated by offshoots as well as seeds. The seedlings grow rapidly and generally begin flowering in the third year. Their greatest growth is in spring and late summer, when they should be watered liberally. In the peak summer months water should be limited, and in winter withheld altogether.

Although *Rebutia krainziana* (1) cannot be mistaken for any other species at first glance, it is possible to observe signs of its kinship to certain other members of the genus. Closely related, in particular, is *R. wessneriana* (2). Some authorities even consider *R. krainziana* to be merely a variety of the former. Its body is flat-topped, globose, up to 8 cm (3 in) across and 7 cm ($2\frac{3}{4}$ in) high, with a greatly depressed, bare crown not enclosed by spines. In mature age it produces profuse offshoots. The skin is

green; only in strong sunlight does it have a violet tinge. Whereas the large, bright red flowers, measuring up to 5.5 cm (2 in) across, are the same as those of *R. krainziana*, the spines are quite different — 2—2.5 cm ($\frac{3}{4}$—1 in) long, and pure white, sometimes with brownish tips.

2

1

A notable characteristic of most high mountain cacti of South America is not only their great adaptability (which is why they are easily grown in collections), but also their ready and early production of brightly coloured flowers. This applies in full measure to the plant named *R. kupperiana* in honour of the German botanist W. Kupper. It is native to Bolivia, where it grows on the plateau of the province of Tarija at elevations of about 2,500 m (8,200 ft). Its body is small, globose at first, later slightly elongated. In mature age it reaches a maximum of 3 cm (1 in) in diameter and 5 cm (2 in) in height. The skin is deep green, acquiring a violet tinge after exposure to direct sunlight. The spines are straight, up to 2 cm ($\frac{3}{4}$ in) long, and brown with a darker tip. Devoid of flowers, it is an inconspicuous plant but during the flowering period it is a veritable gem. The flowers, about 4 cm ($1\frac{1}{2}$ in) across, are composed of very broad, deep red perianth segments. They emerge in spring and then, as a rule, once or even twice more during the summer. They open in the morning for a period of about four days. *R. kupperiana* requires a sunny, adequately ventilated situation, and in winter cool conditions with temperatures not exceeding 12°C (54°F) for long periods. The plants should be watered for the first time in spring, after the buds have formed. If watered too soon in the early stages of bud formation, the buds may develop into offshoots instead, and if watered late they may dry up. This holds true for all species of *Rebutia.*

2

1

Rebutia kupperiana (1) is often found in collections under the name *Aylostera kupperiana.* The genus *Aylostera* was established for species that differ partly in the arrangement of the floral organs, for example with stamens joined to the flower tube. According to the latest opinions, however, the genera *Rebutia* and *Aylostera* cannot be reliably distinguished because of the existence of numerous transitional forms linking the two without a break. Another species likewise often classed under the generic name *Aylostera* is *Rebutia fiebrigii* (2). Like the preceding species it, too, is native to Bolivia, but is found at elevations above 3,600 m (11,800 ft). Its body is slightly prolonged, up to 8 cm (3 in) across, and completely covered with a tangle of straight white spines with pale brown tips. The flowers are about 3.5 cm ($1\frac{1}{3}$ in) across and orange-red.

Rebutia senilis BACKEB.

R. senilis is distributed in various local forms over a large area in the province of Salta in the mountain region of northern Argentina. The most distinctive forms have been described as varieties. Because it is an undemanding, hardy and readily flowering species, it is easily grown even in a window garden, or on a veranda facing south east or south west. It is a globose plant up to 7 cm ($2\frac{3}{4}$ in) across and branching freely at the base. Each areole bears about 25 straight, relatively fine, non-prickly spines up to 3 cm (1 in) long. They are glassy white and completely cover the body, particularly during the resting period when the plant is slightly shrivelled, thus protecting the skin from being scorched by the sun's rays. The allogamous, carmine-red flowers, about 3 cm (1 in) across, arise from the areoles near the base. They emerge in large numbers in early spring and, as a rule, once again in late summer, but in far fewer numbers. *R. senilis* is readily propagated from seeds, which germinate best at a temperature of about 20°C (68°F), as well as from offshoots, which require a slightly higher temperature for rooting. To ensure profuse flowering, provide the plant with a sufficiently nourishing substrate or applications of liquid feed. A further requirement is a sunny, moderately shaded situation and cool, dry conditions in winter.

Rebutia senilis (1) has several varieties, differing not only in general appearance but also in the colour of the flowers. Var. *iseliniana* (2) boasts the largest

2

3

flowers — about 5 cm (2 in) long and up to 4.5 cm (1¾ in) across — coloured orange-red. Var. *sieperdaiana* (3) has yellow flowers, a flat-topped, globose body, and shorter, sparser spines. According to the latest opinions, it is a variety or form of *R. marsoneri.* Besides several other, fully warranted varieties, also often grown is the cactus designated as *R. senilis* var. *kesselringiana* (4), which has yellow, autogamous flowers, a slightly prolonged body and dense yellowish spines. Of late it may also be encountered under the name *R. senilis* ssp. *chrysacantha* f. *kesselringiana.*

4

1

Rebutia violaciflora BACKEB.

Cacti of the genus *Rebutia* generally have yellow, red or orange flowers. Few have flowers of a different colour and so *R. violaciflora*, with its violet blossoms, brings welcome variety to the collection of every cactus grower. The fact that *R. violaciflora* is widespread in cultivation is due not only to the unusual colour plus the ready and early appearance of its flowers, but also to their being self-pollinated, which makes it easy to propagate the plant from seed, even for novice cactus growers. *R. violaciflora* is a miniature, flat-topped, globose plant that produces offshoots but not until a more advanced age. The yellow-brown spines are up to 3 cm (1 in) long, the flowers violet and about 3 cm (1 in) across. Their hue, however, is quite variable in the various populations. Whereas in some it is purplish-violet, in others it may be violet-red to carmine-red. Such plants are designated as var. *knuthiana* or var. *carminea. R. violaciflora* is a high-mountain species that requires ample fresh air in the growing period; during the period of peak summer temperatures, when it undergoes a so-called summer stagnation, it also requires limited application of water. Besides direct sunlight, which poses the danger of sun scorch, it does not tolerate the excessively warm and moist environment of unaired frames and glasshouses. In winter it requires cool conditions and, if kept in a dark place, timely transferral to its summer location so that it may begin to develop buds, which emerge in early spring. It may be put in its summer location in mid-spring.

2

1

Rebutia violaciflora (1) is quite similar but not related to *R. marsoneri* (2). The two can be reliably distinguished when in flower. Whereas *R. violaciflora* has violet blossoms, those of *R. marsoneri* are yellow. Its body is flat-topped and globose, with a markedly depressed crown and green skin; it produces offshoots only very occasionally. The spines are extremely variable, not only in length but also in coloration, which ranges from white through yellow to orange-brown. For this reason also, the often cited varieties cannot be considered as justified. The buds arise in a ring around the base of the stem and are reddish at first. Within a month they develop into deep yellow, allogamous flowers up to 4.5 cm ($1\frac{3}{4}$ in) across. *R. marsoneri* is distributed throughout a relatively large area in the province of Jujuy in northern Argentina.

Schlumbergera truncata (HAW.) MORAN — Christmas Cactus

S. truncata, better known as the Christmas Cactus, is often grown as a house plant. In the late 1820s it was brought to England and described as *Epiphyllum truncatum.* It is native to the state of Rio de Janeiro in Brazil, where it is found at elevations of 900—1,400 m (2,900—4,500 ft). It grows mostly as an epiphyte, i.e. on trunks and branches, very occasionally also on rock faces, on the windward eastern slopes of damp mist forests. *S. truncata* forms shrubs that may measure more than 1 m (3ft) across. Growing from the short woody stem are richly branching, greatly flattened stems resembling leaves. The flowers are zygomorphic, 4—8 cm ($1\frac{1}{2}$—3 in) long, and coloured various shades of reddish-violet. They remain open both day and night for a period of about one week. In their native habitat they are pollinated by hummingbirds. *S. truncata* requires a light, free-draining, nourishing soil with a large proportion of leaf mould, sand and peat. Best suited is a light situation but not in direct sunlight. During the resting period, from August to October, watering should be limited. In winter it forms buds that bloom at Christmas as well as later. At this time it requires warmer conditions, regular watering, and should not be moved, for then the buds might fall or dry up. After the flowering period is over (in early spring) watering should again be limited and the temperature lowered. Springtime is also the best period for repotting. The plant produces new stem segments during the summer and therefore the substrate should be kept continually moist and feed may also be applied. *S. truncata* is readily propagated by broken-off stem segments.

2

Schlumbergera truncata (1), like other cacti, exhibits great variability in its native habitat. Commonest are specimens with pale to deep purple flowers, less common are ones with flowers coloured carmine-red or pale pink to nearly white. The flowers are also variable in the size and shape of the perianth segments. This variability was used in selective breeding to produce

a cultivar with pure white flowers (2).
Besides this, *S. truncata* was also
hybridized with other related species. It
is believed that the specimens most
commonly grown are cultivars and their
hybrids because they flower more
profusely and are hardier from the
cactus grower's viewpoint.

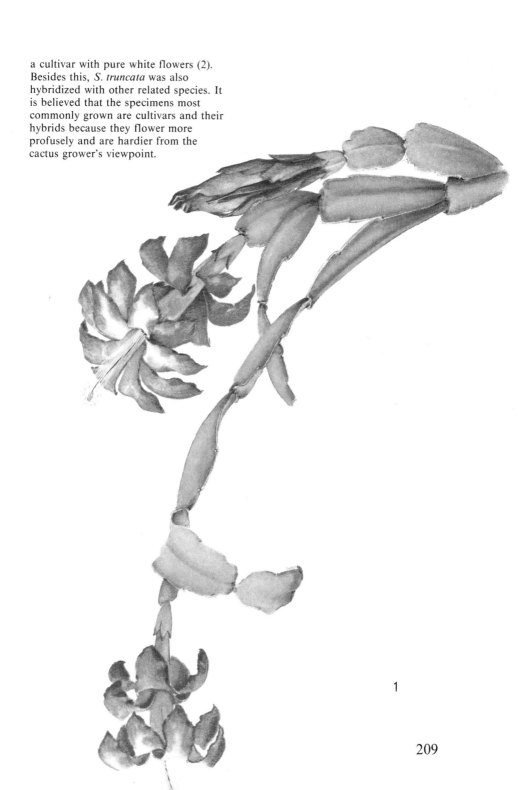

1

Sulcorebutia arenacea (Card.) Ritt.

The genus *Sulcorebutia* includes some 50 species found in Bolivia at elevations of 1,600—3,600 m (5,250—11,800 ft). They are miniature, often clustering, cacti with a stiff body, turnip-like root, pectinate spines, and yellow, red or violet flowers. They flower readily at an early age.

The word *arenacea* means sand-coloured and if you look closely at the plant you will see that that is truly the case. In its native Bolivia it is distributed on the south-western slopes of the Cordillera del Tunari at elevations of about 2,000 m (6,500 ft), where it grows in pure gravel waste without taller vegetation. The small bodies resemble the substrate in which they grow in both shape and colour and are thus very difficult to find when not in flower. The body of *S. arenacea* is flat-topped and globose at first, later regularly globose with a markedly depressed crown. There are six to seven paired spines in each areole; these are pressed close to the body, about 5 mm long, and generally coloured yellow to brown, later turning grey. The flowers are about 3 cm (1 in) across, golden-yellow and allogamous as in all other sulcorebutias. *S. arenacea* grows well on its own roots and flowers readily as a three-year-old seedling. The substrate, however, must resemble that to which it is accustomed in its native habitat, in other words it must be sufficiently free-draining and strongly mineral. In summer the plant should be placed in a sunny, relatively warm spot, and in winter it should be kept at a temperature of 6—12°C (43—54°F).

3

According to the original description, *Sulcorebutia arenacea* (1) is only 2.5—5 cm (1—2 in) across, 2—3.5 cm ($\frac{3}{4}$—1$\frac{1}{3}$ in) high, and often forms clusters. In cultivation, however, it generally does not produce offshoots and only young specimens are of such a small size and flat-topped globose in shape. In mature age the stem becomes slightly prolonged and may measure up to 10 cm (4 in) in diameter. The short, densely arranged spines (2) practically cover the entire skin, thereby protecting the plant against the strong rays of the sun. *S. arenacea*

has several closely related species. One
example is *S. candiae* (3), which forms
clumps and readily produces yellow
flowers from its second year. It, too, is
found in the neighbourhood of the
village of Santa Rosa but at higher
elevations (2,700—2,800 m/ 8,800—9,100
ft). It has longer, slightly twisted spines,
generally coloured yellow, but
sometimes yellow-white or almost
brown.

2

1

Sulcorebutia glomeriseta (Card.) Ritt.

Members of the genus *Sulcorebutia* are perhaps the most ideal cacti for the small window garden or other types of collections with limited space. Many bear brightly coloured flowers before they are 2 cm ($\frac{3}{4}$ in) across, in other words in about their third year. In most cases it is not only their flowers that make them attractive to cactus growers, but also their overall habit. *S. glomeriseta* is one such miniature, readily flowering plant with colourful spines. It is native to the mountain regions of Bolivia. Its body is globose, up to 6 cm ($2\frac{1}{3}$ in) across, and produces numerous offshoots. The areoles are white- to brown-felted and slighly oval in shape. The spines are thin and very numerous, 2—3 cm ($\frac{3}{4}$—1 in) long and coloured white or yellowish, to brownish. The flowers emerging in spring round the perimeter of the plant are small, only 2—3 cm ($\frac{3}{4}$—1 in) across, and coloured pale to dark golden-yellow. This, as well as other members of the genus, are often encountered in collections as grafted specimens that flower profusely but do not retain their natural miniature shape. This is a pity because most can easily be grown on their own roots. Water should be applied when the substrate is completely dry and withheld during the period of peak summer temperatures, when the plants generally undergo a summer dormant period and the roots do not absorb water. Being typical high-mountain cacti, they have greater requirements in terms of sunlight and fresh air. If provided with insufficient sunlight, they lose their typical character, grow too tall, have fewer spines, and flower less profusely. When the grower neglects to provide them with proper ventilation, their skin becomes readily susceptible to sun scorch, which holds doubly true for *S. glomeriseta*. All sulcorebutias require an absolutely dry substrate and a temperature of 5—10°C (41—50°F) in winter.

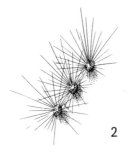

2

Sulcorebutia glomeriseta (1) is not a very typical representative of the genus *Sulcorebutia*. In general appearance it resembles many cacti of the genus *Rebutia* or *Weingartia* and originally, in 1951, it was classed in the genus *Rebutia*. Its reproductive organs, much-branched root system, and radiating spines (2), however, exhibit a greater resemblance to certain species of the genus *Weingartia*, for example

W. multispina (3). *W. multispina* also has numerous bristly, golden-brown spines. The existence of *Sulcorebutia glomeriseta* and *Weingartia*, as well as several other species of the aforesaid genera, has been responsible for the increasingly frequent opinion that the two genera — *Sulcorebutia* and *Weingartia* — should be combined into one.

3

1

Sulcorebutia rauschii FRANK

The spines of *S. rauschii* make it one of the most interesting species of the genus as well as of all South American cacti and that is why it has been intensively cultivated since it was discovered in 1969. Its mass spread in collections has also been due in part to its profuse production of offshoots, which applies in particular to grafted specimens. Like all other sulcorebutias, it is a high-mountain plant of Bolivia. It was discovered in the neighbourhood of the town of Zudáñez in the province of Chuquisaca, at elevations of about 2,700 m (8,800 ft). *S. rauschii* generally forms clusters of numerous bodies about 1.5 cm ($\frac{1}{2}$ in) high and 3 cm (1 in) across, which grow from a turnip-like root. The spines in the oblong areoles are short, only 1—1.5 mm ($\frac{1}{20}$ in) long and curved like claws towards the body. The flowers, produced none too readily, are generally about 3 cm (1 in) across and coloured pinkish-violet to purple. *S. rauschii* is propagated by offshoots that root well, and also by seeds, which are few in number. Grafted specimens produce offshoots almost too profusely, but on the other hand they bear more flowers. Grafted plants as well as seedlings should be placed in a sunny, only slightly shaded spot in summer and provided with sufficient ventilation during the period of peak temperatures even during summer and autumn nights. To prevent deformation of the growing centre at the crown during the winter, the plants should be kept in an absolutely dry substrate at a lower temperature that does not permanently exceed 10—12°C (50—54°F).

3

Sulcorebutia rauschii (1) is just as variable as the other species of this genus, not only in the colour of the skin, which ranges from pale green through silvery green, dark green and brown to violet, but also in the colour of the spines, which may be black as well as golden-yellow (2). Also variable is the shape of the tubercles, the length and

1

number of the spines, and the size of the flowers, which sometimes measure up to 5.5 cm (2 in) across. *S. rauschii* is an interesting plant even without the flowers because of its unusual covering of spines. Each areole bears very short, extremely pectinate spines (3) that are pressed closely against the plant body.

2

Trichocereus candicans (GILL.) BRITT. ET ROSE

Most members of the genus *Trichocereus* are of columnar habit with robust spines and large flowers that open at night. They are very attractive cacti and also excellent stock for grafting. *T. candicans* is a durable stock as well as an attractive plant, differing from the other members of the genus in the fresh green colour of the skin and bright yellow spines. It was originally discovered in the neighbourhood of the town of Mendoza in central Argentina, later also in the province of La Rioja in the Famatina mountains. *T. candicans* remains solitary at first; in mature age it sends out numerous offshoots at the base and forms clusters of many heads. The individual bodies are 12—24 cm ($4\frac{3}{4}$—$9\frac{1}{2}$ in) across and nearly 1 m (3 ft) high. The glossy bright green skin is protected against grazing by animals by the robust, yellow to honey-yellow spines measuring about 8 cm (3 in) in length. The funnel-shaped flowers are very variable in size as well as colour. In some specimens they are only 15 cm (6 in), in others up to 25 cm (10 in) long. They are pure white or pink, violet-red or bright red. Red-flowering plants, intermingled with white-flowering specimens, were described as *T. candicans* f. *rubriflorus. T. candicans* is easy to grow and may be propagated by offshoots as well as seeds. It requires a nourishing substrate and a large container so that its strong, much-branched root system may develop properly. Rapid growth of the bodies may be further promoted by more liberal watering during the growing period. The most suitable location for it in summer is one that is partly shaded and sufficiently airy.

2

Trichocereus candicans (1) does not begin flowering until it is relatively large, but it is attractive even without flowers. The same is true of one of the most familiar of Chilean trichocerei, namely *T. chilensis*, found on the hills of the central Chilean plains in many local forms, each merging smoothly with the next. The most distinctive were described as independent varieties. *T. chilensis* reaches a height of 1—5 m (3—16 ft) and in advanced age branches to a greater or lesser degree in shrub-like fashion. The flowers are

white, 13—15 cm (5—6 in) long, and
slightly closed at night. The fruits are
soft and very tasty. An attractive feature
of this species is the spines, the showiest
being those of var. *borealis* (2). These
are pale grey-brown, greatly thickened,
and the longest measure 8—16 cm
(3—6 $\frac{1}{4}$ in), very occasionally as much as
20 cm (8 in) in length.

1

Weingartia neocumingii Backeb.

W. neocumingii is sometimes encountered in collections under the name of *W. cumingii*. Along with other closely related species, it is distributed in the Rio Grande and Rio Pilcomayo river region in Bolivia, where it grows in stony mountain terrain at elevations of 1,200—2,200 m (3,900—7,700 ft). Its great advantage is that it flowers early and readily, thereby earning the name of 'fuchsia amongst cacti'. It is a solitary plant with a flat-topped globose body becoming slightly prolonged in age and attaining a diameter of about 10 cm (4 in). The flowers are 2.5 cm (1 in) across and coloured an intense yellow to orange. Sometimes two may arise from a single areole. They open in the forenoon for a period of about two days. As a rule they emerge on four-year-old seedlings in a wreath around the crown. Older plants flower profusely and repeatedly throughout the entire growing period. *W. neocumingii* is easy to grow and thus grafting it is quite unnecessary. The only thing it does not tolerate well is lengthy drying out of the substrate during the growing period, as the root system then regenerates with great difficulty and it is hard to stimulate new growth. Under such conditions, as well as at the beginning of the growing period, the skin is readily sun scorched and it must therefore be provided with partial shade. During the stage of full growth, however, it requires a sunny situation with the possibility of frequent airing at peak temperatures, even during summer and autumn nights. The winter temperature should be kept at about 10°C (50°F).

2

3

1

Weingartia neocumingii (1) has straight, needle-like, radiating spines, 0.5—1 cm ($\frac{1}{6}$—$\frac{1}{3}$ in) long. Besides plants with such spines, there are also specimens with much shorter ones, less than 3 mm long (2), designated as var. *brevispina* by C. Backeberg. Very similar to *W. neocumingii* are several other species, such as *W. sucrensis, W. erinacea, W. comarapense* and *W. knizei,* which could all be classed as varieties of *W. neocumingii.* Also related, but readily distinguished, is *W. lanata* (3), named after the prominent cottony areoles (*lanata* means woolly). This plant, like other species that can be distinguished from it only with difficulty (mainly *W. longigibba* and *W. riograndensis*), flowers readily.

INDEX

Numbers in *italics* refer to illustrations

Acanthocalycium glaucum 110—1
   thionanthum 110
   violaceum 110—1
Aporocactus flagelliformis 28—9
Astrophytum capricorne 30—1
   f. *aureum 30*, 31
   cv. *crassispinoides* 30
   var. *maior* 30
   var. *minor* 30
   var. *niveum* 30
   var. *senile* 30
   myriostigma 32—3
      var. *columnare* 33
      var. *jaumavense* 33
      var. *myriostigma* 32
      var. *nudum* 32
      var. *potosinum* 32
      var. *quadricostatum* 33
      var. *strongylogonum* 32
      var. *tamaulipense* 33
      var. *tulense* 33
   niveum 30
   ornatum 34—5
      cv. *niveum* 35
      cv. *virens* 35
Austrocephalus dybowskii 132—3
Aylostera kupperiana 203

Bishop's Cap 32
Blossfeldia fechseri 112
   liliputana 112—3
   minima 112
Borzicactus aureispinus 114—5
   samaipatanus 114—5

Cactus gibbosus 140
Cephalocereus delaetii 44
   senilis 36—7, 44
Cereus jusbertii 104
   latispinus 64
   monstrosus 117
   peruvianus 21, 116—7, 132, 154, 162,
      166, 186
Chamaecereus silvestrii 118—9
   f. *aurea* 118—9
Christmas Cactus 208

Cleistocactus strausii 120—1
Copiapoa cinerea 122—3, 124
   var. *albispina* 123, 124
   var. *cinerea* 123
   var. *columna-alba* 123
   var. *dealbata* 123
   var. *haseltoniana 122*, 123, 124
   krainziana 124—5
      var. *albispina* 124
      var. *haseltoniana* 124
      var. *scopulina* 124, *125*
Coryphantha andreae 40
   bumamma 40—1
   calipensis 38—9
   cornifera 38—9
   elephantidens 40—1
   greenwoodii 40
   sulcolanata 40

Dolichothele longimamma 88

Echinocactus apricus 168
   grusonii 42—3
      f. *alba 42*, 43
Echinocereus albatus 45
   armatus 50
   baileyi 50
   caespitosus 50
   delaetii 44—5
   fitchii 50
   freudenbergeri 44
   knippelianus 46—7
      var. *knippelianus* 46
      var. *kruegeri* 46, *47*
      var. *reyesii* 46—7
   longisetus 44—5
   nivosus 44, 45
   pectinatus 48—9
      var. *rigidissimus* 48
   perbellus 50
   purpureus 50
   reichenbachii 50—1
      var. *baileyi* 50
      var. *bruneispinus* 51
      var. *fitchii* 50
      var. *flavispinus* 51

var. *roseispinus* 51
Echinofossulocactus albatus 52—3
  coptonogonus 54—5
    var. *maior 54,* 55
    multicostatus 56—7
    ochoterenaus 52, 53
    phyllacanthus 56
    vaupelianus 52
Echinopsis aurea 128—9
  eyriesii 21, 96, 108, 126—7
  kermesina 128—9
  mamillosa 128
  oxygona 126—7
  tubiflora 126
  turbinata 126
Encephalocarpus strobiliformis 96—7
Epiphyllum ackermannii 58—9
  anguliger 58, 59
  truncatum 208
Epithelantha micromeris 60—1
  var. *bokei* 60, 61
  var. *greggii* 60
  var. *micromeris* 60, 61
  var. *pachyrhiza* 60
  var. *unguispina* 61
Eriocereus jusbertii 21, 80, 86, 96, 108, 124,
  132, 166
Eriosyce aurata 130, *131*
  ceratistes 130—1
Espostoa haagei 132
  var. *rubrispina* 132
  melanostele 132
  nana 132—3

Ferocactus glaucescens 62—3
  latispinus 64—5
    var. *spiralis* 65
  recurvus 65
  schwarzii 62
Frailea asteroides 114—5
  castanea 114

Glandulicactus uncinatus 66—7
  var. *uncinatus 66,* 67
  var. *wrightii 66,* 67
Golden Barrel Cactus 42
Gymnocalycium achirasense 146
  albispinum 143
  andreae 134—5
  baldianum 134—5
  bayrianum 136—7
  bruchii 142
  cardenasianum 136—7

denudatum 138—9
gibbosum 140—1
  var. *gibbosum* 140
  var. *nigrum* 140—1
  var. *nobile* 141
horridispinum 146
lafaldense 142—3
mihanovichii 144—5
  var. *friedrichii* 144
monvillei 146—7
multiflorum 146
netrelianum 138
saglionis 148—9
  var. *albispinum* 148
  var. *flavispinum* 148
  var. *nigrispinum* 149
  var. *roseispinum* 149
  var. *rubrispinum* 149
schuetzianum 146
spegazzinii 136, 150—1
  var. *maior* 150
uebelmannianum 134
uruguayense 138—9
  var. *roseiflorum* 139

Haageocereus versicolor 120, 121
Hamatocactus setispinus 68—9
  var. *orcuttii* 69
Heliocereus speciosus 28—9, 59
Hildewinteria aureispina 114
Hylocereus undatus 70—1

Leuchtenbergia principis 72—3
Lobivia glauca 153
  jajoiana 152—3
    var. *nigrostoma* 152
    var. *paucicostata* 152—3
  paucicostata 153
  winteriana 155
  wrightiana 154—5
    var. *brevispina* 154
Lophophora diffusa 75
  williamsii 60, 74—5
    var. *decipiens* 74
    var. *jourdaniana* 74
    var. *pentagona* 74
    var. *pluricostata* 74

Mammillaria bella 94
  bocasana 76—7
    var. *flavispina* 76
    var. *multilanata 76,* 77

var. *roseiflora* 77
var. *splendens* 76
bravoae 84
camptotricha 88—9
var. *albescens* 89
candida 78—9
var. *rosea* 78, 79
carmenae 80—1
diguetii 92
geminispina 82—3
var. *nobilis* 82
hahniana 84—5
var. *giseliana* 84
var. *werdermanniana* 84
herrerae 86—7
humboldtii 86—7
longimamma 88—9
mendeliana 84
ortizrubiona 78
parkinsonii 83
pectinifera 90—1
plumosa 80—1
senilis 92—3
solisioides 91
spinosissima 94—5
var. *sanguinea* 94—5
uberiformis 88
woodsii 84, 85
zeilmanniana 77
Mammillopsis senilis 92
Matucana aurantiaca 156—7
currundayensis 156
myriacantha 156, *157*
weberbaueri 156, 157
Melocactus concinnus 158—9
elegans 158
gracilis 158
matanzanus 160—1
seabrasensis 158
Myrtillocactus geometrizans 36—7

Neochilenia floccosa 162, 163
hankeana 162—3
napina 162—3
paucicostata 164—5
var. *viridis* 162
Neoporteria gerocephala 166, *167*
multicolor 166—7
sinilis 166
Notocactus acutus 179
apricus 168—9
var. *nigrispinus* 168

arachnites 185
claviceps 175
concinnus 168—9
crassigibbus 184, 185
graessneri 170—1
var. *albisetus* 170
haselbergii 171
horstii 172—3
var. *purpureiflorus* 172, 173
leninghausii 174—5
f. *apellii* 174
magnificus 176—7
mueller-melchersii 180
ottonis 178—9
var. *vencluianus* 178
purpureus 172—3
rutilans 180—1
var. *longispinus* 180
f. *storianus* 180, 181
scopa 182—3
f. *bruneispinus* 182
f. *candidus* 182
f. *daenikerianus* 182
sucineus 182—3
f. *albispinus* 183
uebelmannianus 184—5
f. *flaviflorus* 184
warasii 177

Obregonia denegrii 96—7
Old Man Cactus 36
Opuntia ficus indica 150
fragilis 98
humifusa 98
microdasys 98—9
var. *albispina* 98—9
'Angel's Wings' 99
var. *rufida* 98
rhodantha 98
tomentosa 150
Oreocereus celsianus 186
trollii 186—7
Oroya borchersii 188, 189
f. *aureotenuispina* 189
var. *fuscata* 189
peruviana 188—9
var. *citriflora* 188
var. *conaicensis* 188
var. *depressa* 188
var. *neoperuviana* 188

Parodia aureispina 192

223

chrysacanthion 196—7
faustiana 194
maassii 190—1
  var. *albescens* 190
  var. *intermedia* 191
mutabilis 192—3
  var. *carneospina* 192
  var. *elegans* 192
  var. *ferruginea* 192
nivosa 194—5
penicillata 196—7
  var. *fulviceps* 196
suprema 191
Pelecyphora pectinata 90
strobiliformis 96—7
Pilosocereus palmeri 100—1

Queen of the Night 102

Rat's Tail Cactus 28
Rebutia albopectinata 199
fiebrigii 202, *203*
heliosa 198—9
krainziana 200—1
kupperiana 202—3
marsoneri 205, *206, 207*
narvaezensis 198, 199
senilis 204—5
  ssp. *chrysacantha* f. *kesselringiana*
    205
  var. *iseliniana* 204
  var. *kesselringiana* 205
  var. *sieperdaiana* 204, 205
violaciflora 206—7
  var. *carminea* 206
  var. *knuthiana* 206
wessneriana 200, *201*

Schlumbergera truncata 208—9
Selenicereus grandiflorus 102—3
Solisia pectinata 90

Sulcorebutia arenacea 210—1
candiae 210, 211
glomeriseta 212—3
rauschii 214—5

Thelocactus bicolor 104—5
  var. *bicolor* 104
  var. *bolansis* 104
  var. *flavidispinus* 104
  var. *schottii* 104
heterochromus 104, *105*
hexaedrophorus 106—7
  var. *decipiens* 106, *107*
schwarzii 104
wagnerianus 104
Trichocereus candicans 216—7
  f. *rubriflorus* 216
chilensis 216
  var. *borealis* 217
Turbinicarpus pseudopectinatus 90, 91
schmiedickeanus 108—9
  var. *dickinsoniae* 108
  var. *flaviflorus* 108
  var. *gracilis* 108
  var. *klinkerianus* 108—9
  var. *macrochele* 108, *109*
  var. *schmiedickeanus* 108
  var. *schwarzii* 109

Weingartia comarapense 219
cumingii 218
erinacea 219
knizei 219
lanata 218, 219
longigibba 219
multispina 213
neocumingii 218—9
  var. *brevispina* 219
riograndensis 219
sucrensis 219